GIS and Remote Sensing Techniques in Land- and Water-management

GIS and Remote Sensing Techniques in Land- and Water-management

Edited by

A. van Dijk

DHV Consultants,
Amersfoort, The Netherlands

and

M.G. Bos

International Institute for Land Reclamation and Improvement (ILRI),
Wageningen, The Netherlands

KLUWER ACADEMIC PUBLISHERS
DORDRECHT / BOSTON / LONDON

Library of Congress Cataloging-in-Publication Data

ISBN-13:978-94-010-6492-7 e-ISBN-13:978-94-009-0005-9
DOI: 10.1007/978-94-009-0005-9

Published by Kluwer Academic Publishers,
P.O. Box 17, 3300 AA Dordrecht, The Netherlands.

Sold and distributed in North, Central and South America
by Kluwer Academic Publishers,
101 Philip Drive, Norwell, MA 02061, U.S.A.

In all other countries, sold and distributed
by Kluwer Academic Publishers,
P.O. Box 322, 3300 AH Dordrecht, The Netherlands.

Printed on acid-free paper

Table of Contents

Preface

The topic of Fifth ICID-NL day (the Dutch National Committee of the International Commission on Irrigation and Drainage) was "GIS and Remote Sensing Techniques in Land- and Water management. The Fifth ICID-NL day was held at DHV headquarters on the 25th of March 1999.

Managing land and water is a complex affair. Constantly man has to take decisions to allocate and use the natural resources. Decision and action in any use of resources often have strong interactions and side effects on others. Therefore it is extremely important to monitor and forecast the impacts of the decisions very carefully, so that all actions leading to achievement of the goal are attuned to one another and negative impacts are mitigated. Compulsory for monitoring and forecasting are **reliable information** and **clear data manipulation procedures**.

Land and water management over the world is quite different. This can be caused by the user requirements and the climatological, geographical, social and political conditions. But in nearly all situations **reliable information** is missing and **clear data manipulation procedures** are lacking.

The Remote Sensing (RS) and Geographic Information System (GIS) techniques have advanced considerably during the last 30 years. RS has a considerable potential to provide reliable data. GIS is an easy tool to manipulate and analyse the data in a clear and fast way. The costs of remote sensing data and interpretation have come down dramatically by the reduction of software and hardware prices in the world of information technology. A personal computer with simple and cheap raster and vector map software has nowadays sufficient capacity to analyse complex satellite images. Some images are now freely available on internet.

Since Remote Sensing (RS) and Geographic Information System (GIS) are extensively discussed in the proceedings, additional information about these subjects are provided here.

Reliable information is actual and discriminatory, it has an acceptable level of accuracy and is obtained in a timely, cost-effective manner. Information on land and water can be acquired according to the conventional way: by sending survey teams into the field. However fieldwork is often handicapped by considerations of time, distance, weather and the size, diversity and inaccessibility of the survey areas. Therefore collecting information according to the conventional way can be inefficient. Inventories to collect information normally make intensive use of **Remote Sensing (RS)**. Remote sensing is the gathering and processing of information about the earth's environment, particularly its natural, agricultural and water resources, using photographic and related data acquired from an aircraft or satellite. Remote sensing techniques are proposed because they are often the only way to obtain accurate information economically and quickly.

Although remote sensing data interpretation makes an inventory easier, there are limitations, which must be recognized. The degree to which cover types can be recognized depends on the type, quality, scale and season of the remote sensing data collection. A remote sensing interpretation has to be followed by field checks. Remote sensing techniques are therefore used to improve and reduce field work rather than to take its place.

The following characteristics of the high resolution satellite data have an important bearing on the natural resources studies.

- Synoptic view
  The high resolution satellites cover areas of about 185 by 185 kilometres (TM) or 60 by 60 kilometres (SPOT) and have a ground resolution of 30 meters (TM), 20 meter (SPOT) or 1 meter (IKONOS). These characteristics allow for surveys over a large area at scales from 1:250 000 to 1: 10 000. Important features such as the influence of climate, parent material, topography and soils on vegetation can be inferred over large areas.
- Repetitive coverage
  The satellite passes over the same area every 16 days. The repeated coverage provides an unequalled opportunity for scene selection for the time best suited for the natural resources investigations.
- Multi-spectral capabilities
  Satellite data is recorded in several distinct parts of the spectrum. These multi-spectral capabilities have proven to be extremely useful in land and water inventories.
- Near-orthographic projection
  The almost uniform vertical projection and the advanced geometric correction methods mean that satellite data can be regarded as having topographic specifications.

All this means that the high resolution satellite data is a very inexpensive tool for the data collection of natural resources.

Data manipulation is the management and analysis of spatially referenced information in a problem solving synthesis. Until some few years ago spatially referenced information was only stored in analogue form: reports, maps, graphs and tables. Information can have different formats, scales, levels of detail. These different forms of analogue information make data manipulation a cumbersome affair.

A **Geographical Information System (GIS)** allows the manipulation of the data by modern methods of computer data processing, archiving and retrieval techniques. A GIS consists of a data bank and a system of overlay techniques, permitting storing, checking, integrating, analyzing and displaying of spatial data.

Although a GIS makes the data manipulation easier, there are limitations that must be recognized. The results of the GIS data manipulation depend in many ways on the quality of the data, the format to be used for the exchange of the data and the resolution of the data. GIS cannot recover errors in the original data files (many inaccuracies will not be noticed in nice looking maps). Errors can be introduced during digitizing, gathering, converting, storing, processing, integrating and overlaying of the data. Also the manipulation of spatial distribution of features into homogeneous mapping units can cause a considerable loss of information. The data overlaying can result in a summation of the errors. Furthermore the analysis of informational data with a high variance can introduce mistakes. An important consideration is the law of the limiting accuracies: the data set with the lowest accuracy determines the accuracy of the new data set after data manipulation.

Combination of RS and GIS into an RS/GIS system. The role of RS and GIS is fundamentally simple. RS is merely to provide information, GIS to handle information. The functions of RS and GIS suggest a union of RS and GIS. This union becomes more realistic when one realizes that GIS information also can be used to improve the classification results of an RS image. For instance: the classification of vegetation in a mountainous area. In this environment the altitude influences the types of vegetation. The contour lines and the data manipulation in the GIS will result in a better classification of the vegetation using an RS image.

Amersfoort, April 2000
A. van Dijk
M.G. Bos

1. Why would we use a GIS database and remote sensing in irrigation management?

MARINUS G. BOS
International Institute for Land Reclamation and Improvement
P.O. Box 45, 6700 AA Wageningen, The Netherlands
<m.g.bos@ilri.agro.nl>

Abstract

In irrigated agriculture the future goal will be to use (natural) resources more effectively. This goal can be re-worded as; improve the performance of irrigated agriculture. To facilitate this goal a set of up-to-date tools are needed. The most prominent new tool is named "performance assessment". Performance assessment can be defined as the systematic observation, documentation and interpretation of the management of a project with the objective of ensuring that the input of resources, water delivery schedules, targeted outputs and required actions are proceeding according to plan.

A systematic and timely flow of actual (measured or collected) data on key aspects of a project is a sine qua non for the monitoring of performance to become an effective management tool. This data should contain sufficient information for the managers to answer questions effectively. The use of a GIS database is the next tool needed for this purpose.

Filling a database with measured and collected data may cost a considerable part of the (often limited) management budget. To make this process feasible, modern technology must be used. Remote sensing is a promising tool to measure a variety of data in a cost-effective manner.

1. Improving the performance of irrigation

Irrigated agriculture refers to a range of phenomena. At one end it refers to the supply of water to grow a crop, at the other end it refers to the development of institutions and the related laws or relationships in a society. In this paper, irrigated agriculture is defined as:

> *The process by which individual water users, user organisations, and irrigation management institutions use water to grow a crop in relation to their goals, other resources and the environment.*

A. van Dijk and M. G. Bos (eds.), GIS and Remote Sensing Techniques in Land- and Water-management, 1–8.

This definition distinguishes between three levels of 'institutions'; individuals, organisations and institutions. Performance assessment is tool to achieve effective project performance by providing relevant feedback to project management at all three levels. As such it may assist in determining whether performance is satisfactory and, if not, which corrective actions need to be taken in order to remedy the situation.

Performance assessment can be defined as the systematic observation, documentation and interpretation of the management of an institution with the objective of ensuring that the input of resources, targeted outputs and required actions are proceeding according to plan. The ultimate purpose of performance assessment is to achieve efficient and effective performance of the institution by providing relevant feedback to management at all levels.

A systematic and timely flow of actual (measured or collected) data on key aspects of the institution is a sine qua non for the monitoring of performance to become an effective management tool. This data should contain sufficient information for the managers to answer two simple questions:

> ***"Are we doing things right?"***
> *Which asks whether the intended level of service that has been set (or agreed upon) for the institution is being achieved.*
> *This is the basis for good* ***'operational performance'***.

and

> ***"Are we doing the right thing?"***
> *A question that aims at finding out whether the wider objectives of the institution are being fulfilled, and fulfilled efficiently.*
> *The latter is part of the process of assessment of* ***'strategic performance'***.

Operational performance is concerned with the routine implementation of the agreed (or pre-set) level of service. It specifically measures the extent to which intentions are being met at any moment in time, and thus requires that actual inputs of resources and the related outputs be quantified on a regular basis.

Strategic Performance is a longer term activity that assesses the extent to which all available resources have been utilised to achieve the service level efficiently, and whether achieving this service also meets the broader set of objectives. Time-series of the indicator and its rate of change commonly are used in this activity.

Available resources in this context refers not merely to financial resources: it also covers the natural resource base (water and infrastructure) and the human resources provided to operate, maintain and manage 'water systems'. Strategic management involves not only the water system manager, but also higher level staff in agencies and at national planning and policy level.

A simplified flow chart, indicating the inter-relationship between strategic, operational and diagnostic management activities is shown in Figure 1.

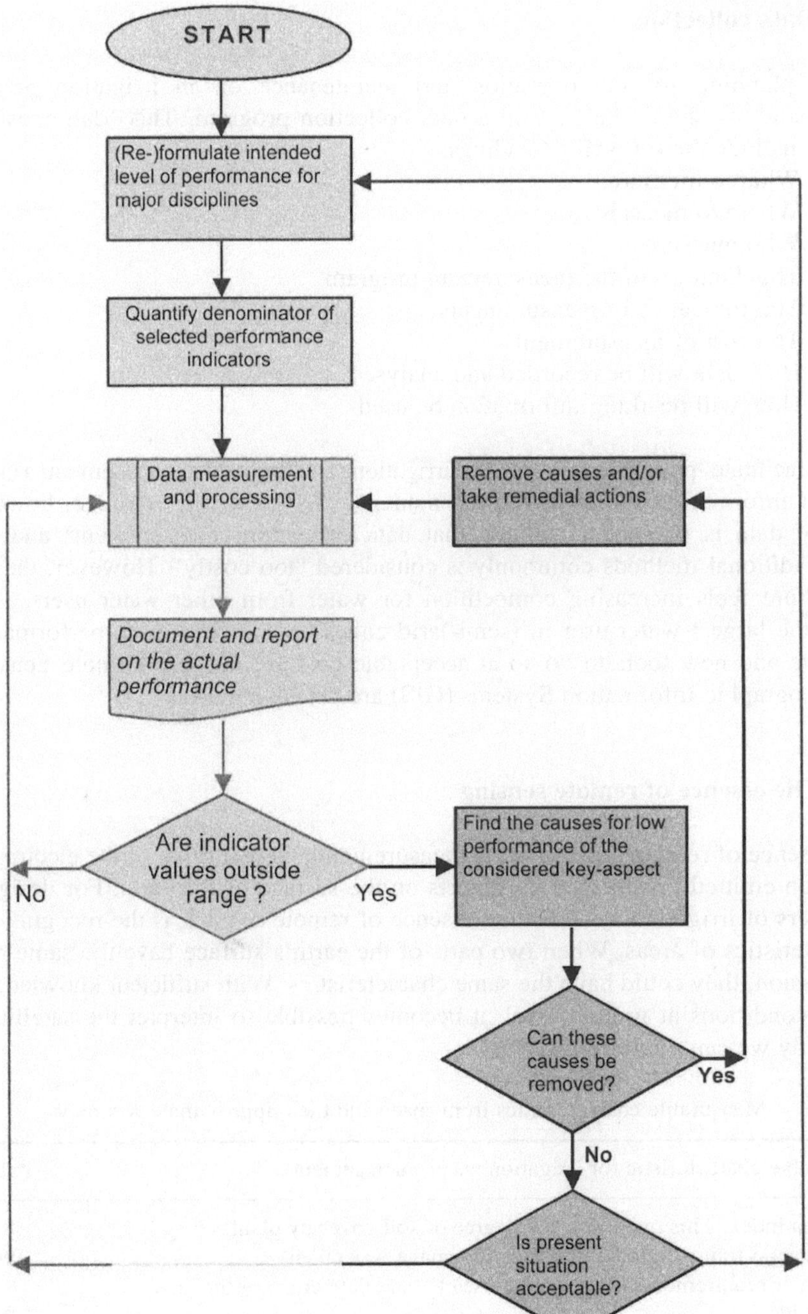

Figure 1. Schematic inter-relationship between strategic (light grey), operational (grey) and diagnostic (dark grey) management activities

3. Data collection

While planning for the operation and maintenance of an irrigation project, due considerations should be give to a data collection program. This 'due consideration' should include the following headings:

- What to measure
- Where to measure
- Who measures
- The duration of the measurement program
- The frequency of measurements
- The cost of measurement
- How data will be recorded and analysed
- How will resulting information be used

To facilitate 'performance oriented' irrigation (and drainage) management, a continuous flow of information should reach the manager. As shown in Figure 2, however, this flow of data is of such a volume that data collection (measurement) and handling with traditional methods commonly is considered "too costly". However, the irrigated agriculture feels increasing competition for water from other water users. Irrigation, being the largest water user in (semi-)arid zones has no choice; its performance must improve and new tools to do so at acceptable cost are needed. Remote Sensing (RS) and Geographic Information Systems (GIS) are promising tools.

4. The essence of remote sensing

The essence of remote sensing is the measurement and recording of the electromagnetic radiation emitted or reflected by objects on the surface of the earth. For designers and managers of irrigation systems, the essence of remote sensing is the recognition of the characteristics of areas. When two parts of the earth's surface have the same reflection or emission, they could have the same characteristics. With sufficient knowledge on the actual conditions at ground level, it becomes possible to interpret the satellite image. Currently we can view from space:

Table 1. Measurable characteristics from space and their approximate accuracy

Measurable characteristic for irrigation water management	Accuracy
Leaf area index. This quantifies the degree of soil cover by plants	85%
Actual evapo-transpiration at the time the image was taken	90%
Crop water requirements. This can be used to calculate crop coefficients	85%
Crop yields	80%
Land wetness; can be used to quantify uniformity of water delivery or to determine the need for artificial drainage	75%
Salinity of the top soil. To observe salinity, the image must be taken at the end of a dry period without irrigation water supply	40%

The cost of data collection and analysis on the above characteristics is low if the irrigated area is sufficiently large (say greater than about 10 000 ha). Bastiaanssen et al. (1999) reports a cost of 0.02 US$ per ha for a 14.000 ha project in India.

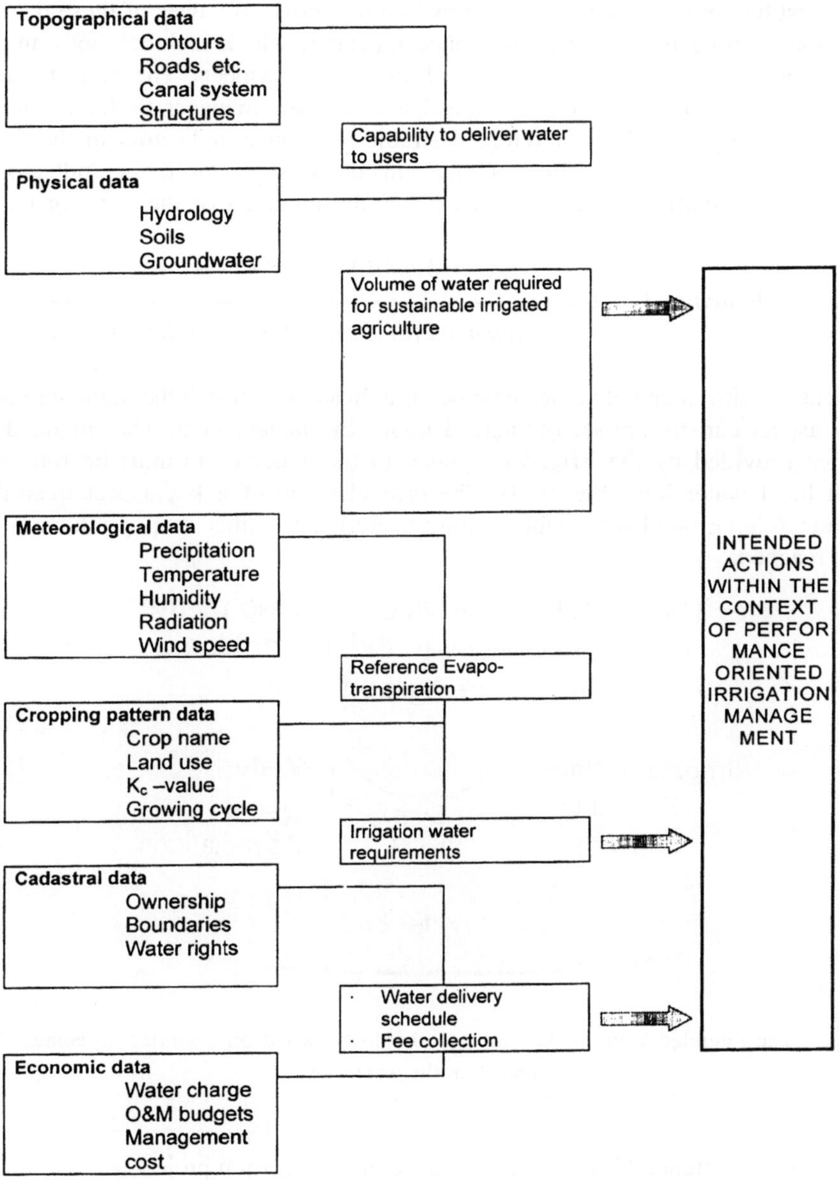

Figure 2. Relation between data and irrigation management.

5. General Shape of an Indicator

The GIS database should be used to determine the degree of satisfaction on a key aspect under consideration. Hence, the actual (measured or collected) data on key aspects of a project must be compared with intended or limiting (critical) values of these data. This comparison is done through a performance indicator, which includes both an actual value and an intended (or critical) value of data on key aspects. The indicator further should be accompanied with information that allows the manager to determine if the deviation is acceptable. It is therefore desirable to express indicators in the form of a dimensionless ratio. Depending on the considered key aspect, the indicator thus is the ratio of actually measured versus either the intended or the limiting (critical) situation. Hence:

$$\textit{Performance Indicator Value} = \frac{\textit{Actual Value of Key Aspect}}{\textit{Intended (or Critical) Value of Key Aspect}}$$

Indicators with intended values include all indicators of which the intended value of the key aspect can be pre-set (or agreed upon) by management. The intended level of service provided by the irrigation agency to the water users must be founded on a sound local water law (Figure 3). The critical value of a key-aspect quantifies a key-aspect (e.g. groundwater table, salinity, etc.) that limits crop production if its value is reached.

Figure 3. The intended value of key aspects should be based on a service agreement that is founded on the water law.

Setting of the intended level of performance of an irrigation project (or institution) is recommended to be done from at least the tree major fields of activities:

- Water balance management
- Environmental management, and
- The socio-economics of project management

Under these three headings we recommend to select performance indicators from the long-list as recommended by the International Commission on Irrigation and Drainage (Bos 1997). The following of these indicators contain a 'measurable' that can be quantified by remote sensing:

Name of Indicator	Component which is measurable with RS	Remarks
Field application ratio	Actual evapo-transpiration, ET_a	Using the copping pattern and software like CROPWAT and CRIWAR, the ET can be extrapolated over the season
Overall consumed ratio	Actual evapo-transpiration, ET_a	This is one of the basic recommended indicators
Sustainability of irrigated area	Actual cropping pattern	If averaged over the year, this indicator quantifies the cropping intensity of the irrigated area
Relative groundwater depth	Actual depth to groundwater	Only in case of water logging during wet periods
Relative salinity	Leaf area index and salt at soil surface	RS image should be taken at end of canal closure period
Yield versus water cost ratio	Crop yield	RS image to be taken during characteristic growth stage of crop
Yield versus water supply ratio	Crop yield	As above

As illustrated in Figure 1, a considerable number of data needs to be measured and handled. Subsequently, the results should be reported (and presented) in such a way that management decisions can be supported. A GIS is a powerful tool to do the required calculations (with the measured data) and to present results as a function of time (e.g. salinity) or space (e.g. water delivery).

6. References

Bos, M.G. 1997. Performance Indicators for Irrigation and Drainage. *Irrigation and Drainage Systems*, Kluwer, Dordrecht, Vol. 11, No. 2, pp. 119-137.

Bos, M.G., D.H. Murray-Rust, D.J. Merrey, H.G. Johnson and W.B. Snellen, 1993. Methodologies for assessing performance of irrigation and drainage management. Irrigation and Drainage Systems, Kluwer Academic Publishers, Dordrecht, Vol.7, No.4.

Bos, M.G. and J. Nugteren. 1990. On irrigation efficiencies. 4th edition [1st edition 1974]. ILRI Publication 19. Wageningen: International Institute for Land Reclamation and Improvement. pp 141.

Bos, M.G., W. Wolters, A. Drovandi and J.A. Morabito. 1991. The Viejo Retamo secondary canal — Performance evaluation case study: Mendoza, Argentina. *Irrigation and Drainage Systems* 5: 77-88.

Bos, M.G., J. Vos and R.A. Feddes. 1996. CRIWAR 2.0; A simulation model on crop irrigation water requirements, ILRI Publication 46. Wageningen: International Institute for Land Reclamation and Improvement. pp 117.

International Commission on Irrigation and Drainage (ICID). 1978. Standards for the calculation of irrigation efficiencies. *ICID Bulletin* 27(1): 91-101.

International Irrigation Management Institute (IIMI). 1989. Efficient irrigation management and system turnover. ADB Technical Assistance TA 937-INO, Indonesia. Final Report. Volume 2.

Murray-Rust, D.H. and W.B. Snellen. 1993. Irrigation system performance assessment and diagnosis. (Joint publication of IIMI, ILRI, and IHE). Colombo, Sri Lanka: IIMI.

Small, L.E. and M. Svendsen. 1990. A Framework for assessing irrigation performance. *Irrigation and drainage systems*, Kluwer, Vol.4, No.4, pp 283-312. Revised edition as: Working Paper on Irrigation Performance 1. Washington, D.C.: International Food Policy Research Institute.

Smit, M. 1989. CROPWAT; Program to calculate irrigation requirements and generate irrigation schedules, Irrigation and Drainage Paper 46, FAO, Rome, pp 133.

Wolters, W. 1992. Influences on the efficiency of irrigation water use. ILRI Publication No. 51. Wageningen: International Institute for Land Reclamation and Improvement.

2. Can WARMAP save the Aral Sea?

ALBERT VAN DIJK *and* ROB DEN HAAN
DHV Consultants
P.O. Box 1399, 3800 BJ Amersfoort, The Netherlands
<geo@cons.dhv.nl>

Abstract

We remember the Aral Sea as one the world's worst ecological disasters. The volume of the Aral Sea has reduced dramatically; from 1100 km^3 in 1960 to only 270 km^3 in 1995. Salinization has been one of the worst effects. The water-table has risen, often bringing the groundwater into the rooting zone of the irrigated crop. Foreign aid programmes make it possible to implement high-tech Information Technology systems. One of the projects under the Tacis programme is the "WAter Resources Management and Agricultural Production in the Central Asian Republics" (WARMAP). Such projects will certainly help to solve water management problems and balance the different uses of water. But issues like maintenance and management will remain. The maintenance of the irrigation system is poor. Canals are broken and choked with reed and dirt. The top-down centralized management system is unsuited to successful development of new structures. It is focussed on technology, unable to redirect human resources and lacking in experience of market-driven processes.

1. INTRODUCTION

For most of us, mention of the Aral Sea conjures up images of camels nibbling sparse desert vegetation and scratching their backs against rusting fishing boats, left high and dry by the receding waters. We recall maps and figures showing the Aral Sea's dramatic reduction in volume; from 1100 km^3 in 1960 to only 270 km^3 in 1995. We remember the Aral Sea as one the world's worst ecological disasters.

Few people understand exactly what has happened. The facts are these: in Central Asia, two mighty rivers — the Amudarya and the Syrdarya — flow eastwards from the foothills of the Himalayas to the land-locked Aral Sea. Together they transport 110 billion cubic metres of water a per year, one and a half times the annual discharge of the Rhine and half times that of the Indus. Irrigation has long been practised in Central Asia, but it was always on a small scale, leaving plenty of water to flow into the Aral Sea and balance losses due to evapotranspiration.

All this changed after World War II when the irrigated area was doubled to 7.9 million ha and the Soviet regime decided that cotton was the ideal crop, providing employment for millions of local people and contributing heavily to the economy.

A. van Dijk and M. G. Bos (eds.), GIS and Remote Sensing Techniques in Land- and Water-management, 9–15.

The effect on the Aral Sea was ignored. Inflow was cut from 56 billion cubic metres in the fifties to only 2 billion cubic metres in the eighties. Worse still, the remaining water was saline and loaded with fertilizers, insecticides and pesticides.

The effects were dramatic. The Sea shrank to less than half its original size. Almost all its flora and fauna were destroyed. Thousands of fishermen lost their jobs. The tempering effect of the Sea on the local microclimate was lost: winters became colder and summers hotter. Salt storms became common, threatening the health of local people.

Salinization has been one of the worst effects. The water-table has risen, often bringing the groundwater into the rooting zone of the irrigated crop. In these semi-arid areas, the groundwater contains a high concentration of fossil salts. These salts inhibit cotton growth. Yields have declined from 3.8 tons per ha in 1980 to 2 tons now. Salinization currently affects 75% of the irrigated land in the area and the World Bank estimates the economic damage at US$ 600 million per year.

In 1993, to manage these problems and coordinate foreign aid, the five Central Asian states (Kazakhstan, Kyrgyzstan, Tajikistan, Turkmenistan and Uzbekistan) established an "Executive Committee of International Funds for Saving the Aral Sea" (EC-IFAS). The Funds comprise contributions from governments in the region, international donors like the World Bank and the EU, and bilateral donors like the Netherlands and US governments.

The five Central Asian Republics (CARs) intend to use EC-IFAS to achieve a 15% reduction in water use. Foreign aid programmes make it possible to implement high-tech Information Technology and Telecommunication systems and the CARs believe that these can solve all the problems.

2. WARMAP

In 1992 the five CARs asked the European Union for help with the Aral Sea. The EU quickly appreciated that basin-wide problems were involved. It prepared a project under the Tacis programme focusing on land and water management in the entire basin: the "Water Resources Management and Agricultural Production in the Central Asian Republics" (WARMAP) project.

The first phase (1955 to 1998) comprised technical assistance to EC-IFAS. The second started in February 1998 and its general aims are:

- to strengthen national and regional planning in the area of land and water management (building on and extending the work done in WARMAP 1) through staff training and the development and application of planning systems;
- to promote an economic approach to land and water management by coordinating national efforts and external aid and by providing short-term economic and legal policy advice and reliable data on relevant subjects.

More specific objectives are:

- to develop the planning and management capacities of recipient staff;
- to prepare water use, water planning and other legal interstate agreements;

- to establish management information systems.

The following deliverables have been agreed with the EU and beneficiaries:

- normative and legal documents (interstate agreements) to improve water division and the sustainability of water resources management at regional level;
- a regional information system designed to support land and water resources management (WARMIS) at regional level;
- analyses of data from a Water Use and Farm Management Survey (WUFMAS) identifying ways to increase farm productivity and save water;
- technical assistance to EC-IFAS for the management of the "Water and Environmental Management Project for the Aral Sea Basin", sponsored by the Global Environmental Facility (GEF/IBRD).

3. WUFMAS: reliable current data

The Water Use and Farm Management Survey (WUFMAS) began in early 1996. The aims are:

- to provide accurate current information about farm production and use of resources (land, water, labour, machinery, fertilizer, agrochemicals, seed and capital) on individual farms.
- to calculate financial and economic returns on water and other inputs (e.g., fertilizers, pesticides).
- to estimate regional and local differences in farm production and productivity.
- to facilitate strategic and investment planning within agriculture at national and regional level.
- to provide baseline information for the identification of possible pilot projects on improved farm management, profitability and water use.

The results of the WUFMAS survey will be used as input for the WARMIS decision support models and for studies of possible water-saving measures at farm-level in the GEF project.

4. WARMIS: a high-tech information system

WARMIS is a system for the collection, storage, processing and analysis of data about the past and present situation and use of land and water resources. It has two aims: to help strengthen national and regional planning within the Aral Sea basin, and to promote an economic approach to land and water management by providing reliable data and analysis tools.

Its main tasks are:

- to establish the present structure of water resources and use;
- to establish past water distribution patterns and the demands that dictated them;
- to provide basic data for use in economic analysis of regional issues;

- to provide programmes (models) for first-level analysis of regional issues;
- to supply information and analyses as a basis for regional agreements;
- to establish regular liaison and information exchange between participating organizations.

The main regional water issues to be analyzed are:

- distribution of available water between the five republics: a complicated issue since many watercourses cross and re-cross boundaries.
- analysis of surface and groundwater salinization. Agricultural production upstream is reduced by increased (secondary) groundwater salinity, whereas downstream water quality is reduced by saline drainage water from the upstream areas.
- improvement of water management in irrigation networks and on farms. The water-table can be lowered and secondary salinization reduced by cutting water demand and increasing drainage flow.
- balancing the different needs and resources of the republics involved; e.g. upstream countries (Kyrgyzstan, Tajikistan) release water in the winter to generate power, whereas downstream countries need water during the summer to irrigate crops. Kyrgyzstan and Tajikistan have few natural resources, whereas Uzbekistan and Turkmenistan have major oil and gas deposits, as well as an overcapacity in thermal power stations.

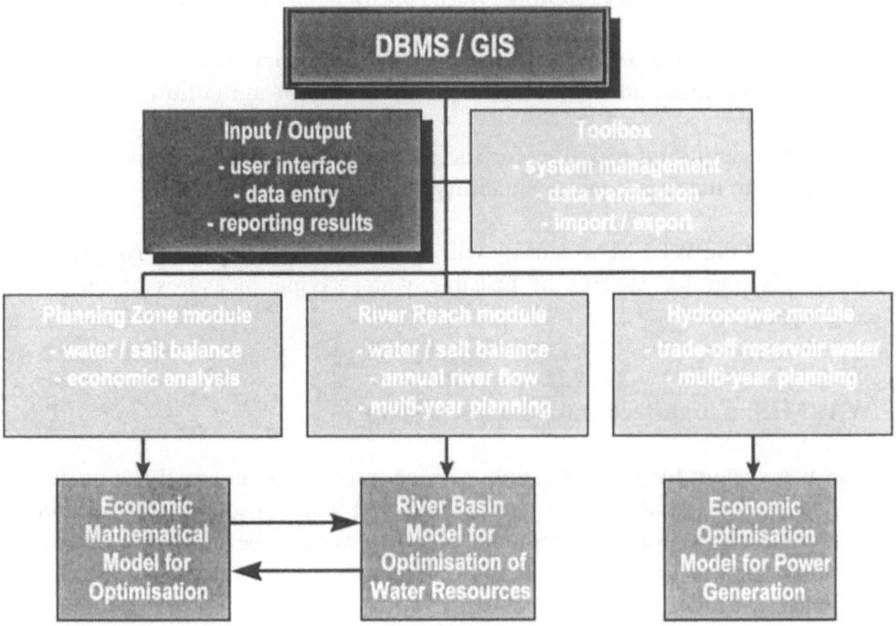

Figure 1. Components of WARMIS.

WARMIS comprises:

- a Database Management System (DBMS), containing tabular data;
- a Geographic Information System (GIS), containing spatial data and spatial analysis tools;
- a User Interface for data input and output;
- a Toolbox for system maintenance, data verification, exchange and security, user authorization, etc. Given the previous Soviet reliance on planned data, data checking and validation are particularly important. WARMAP-2 has developed special tools for validation and graphic displays for data analysis, identifying missing data, outliers or falsifications;
- three strategic analysis and/or decision support modules. Each module consists of one or more mathematical models able to analyze historical data or simulate scenarios for future planning. The modules with their models may be used independently or together. The three modules are:

4.1. Planning zone module

This is designed for socio-economic evaluation of water and land use policies at planning zone level, within the constraints set at national and/or regional levels. Two mathematical models are being developed:

- a water and salt balance model.
- an economic optimization model to identify improvements with and without various constraints. The model will work together with the river basin model (see below), constantly using feed-back on water availability and water quality as input for the next iteration.

4.2. River reach module

Three models are being developed to analyze river water quantity and quality:

- a water and salt balance model for each river reach. Used together with the economic optimization model, this will model the consequences at planning zone level.
- a river basin model to simulate water availability and demand in terms of both quantity and quality. It will suggest (regional) constraints for economic sub-optimizations at national and zone levels. The model will analyze the outcome of the planning zone economic optimization model to see whether constraints are met, and will strive to optimize water use. If constraints cannot be met, the analysis results will be used as new input for the next iteration of the economic optimization model.
- an annual flow model for the Amudarya will support the management and distribution of river water in accordance with agreements and (national) water demand. A similar model will be developed for the Syrdarya with the assistance of the USAID EPIC programme. This will be closely connected with the Hydropower Module since the Syrdarya feeds a series of hydropower stations.

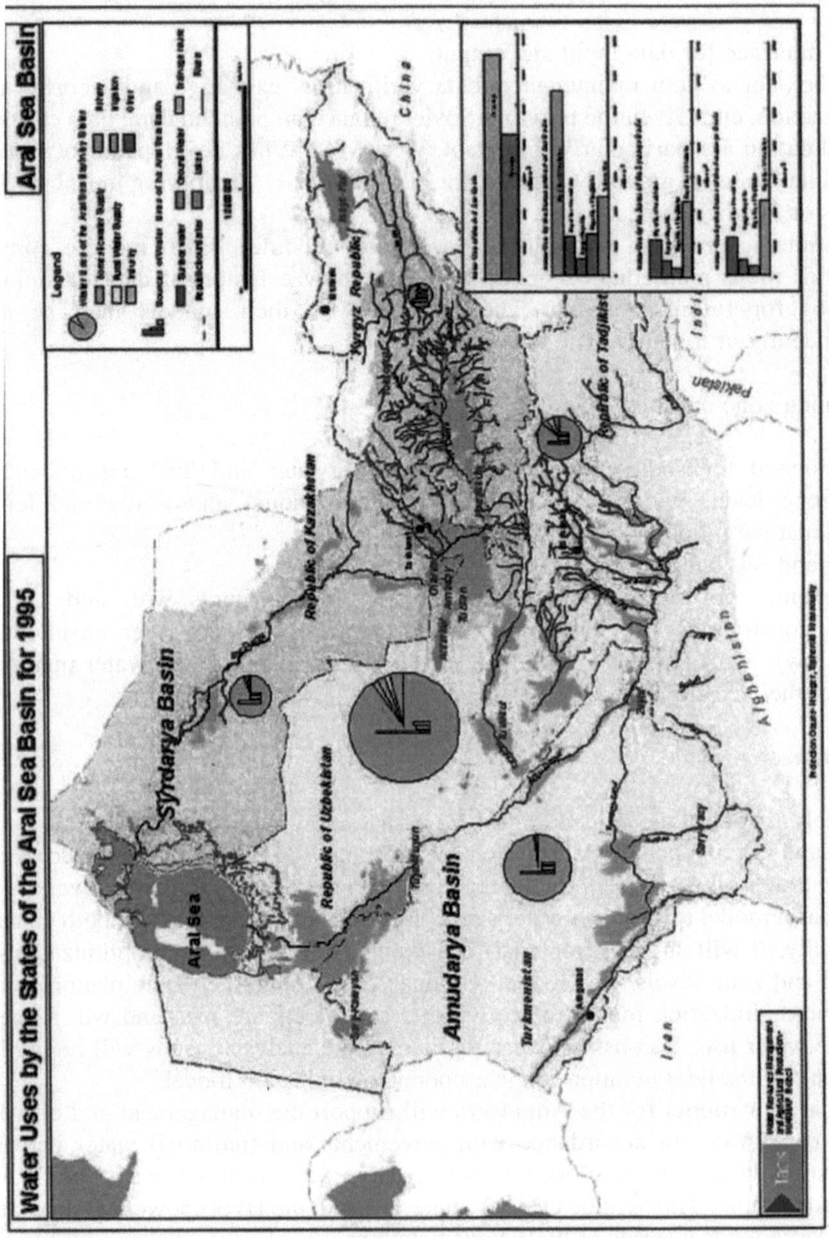

Figure 2. Aral Sea Basin and use of water by States.

4.3. Hydropower module

This module will evaluate economic trade-offs achievable by optimizing the use of stored water for power generation and irrigation. For example, if Uzbekistan reduces the need for hydropower by providing Kyrgyzstan with oil and gas in winter, more water will be available for its agriculture in the summer. A sub-database is being developed containing information on the comparative costs of power generation and on total year-round and region-wide demand for power.

The DBMS used by WARMIS is MicroSoft Access 2.0, with applications written in either Access or Visual Basic. The Geographic Information System used by WARMIS is ArcView, with ArcInfo being used for development and special tasks only. The spatial data within the WARMIS GIS are all related to the tabular database described in the previous section. ArcView can be seen as the DBMS for the spatial data within WARMIS, and also serves as an interface which allows the user to access tabular data by pointing at objects on a map (e.g. in the case of the river reach water/salt balance). Further results from data verification and data analysis within the 3 modules can be presented in a spatial context; e.g. thematic maps or network schemes.

5. Conclusions

High-tech WARMAP Information Technology and Telecommunication systems will certainly help to solve water management problems and balance the different uses of water. But they will not solve all the problems. Issues like maintenance and management will remain. The maintenance of the irrigation system is poor. Canals are broken and choked with reed and dirt. The top-down centralized management system is unsuited to successful development of new structures. It is focused on technology, unable to redirect human resources and lacking in experience of market-driven processes.

The Aral Sea cannot be saved. But there are other possibilities. For part of the year, the Amudarya carries good quality water to the old Aral Sea. This water currently evaporates or sinks into the ground. It could, however, be retained by dams and sluices to restore the wetlands of the Amudarya. An Euroconsult study suggests that this might cover an area of about 50 by 40 km. As such, it could support fishermen, safeguard biological diversity and provide a valuable rest area for migratory birds.

6. References

Dijk, van Albert, Haan, den Rob, Can Warmap save the Aral Sea?, Land & Water International 93 (1999) p. 6-9.

3. Remote sensing for inland water quality detection and monitoring: State-of-the-art application in Friesland waters

ARNOLD DEKKER[1], STEEF PETERS[2],
ROBERT VOS[2] *and* MACHTELD RIJKEBOER[2]

[1] *Dr. Arnold G. Dekker, CSIRO Land and Water, GPO Box 1666, ACT 2601 Australia <Arnold.dekker@cbr.clw.csiro.au>*

[2] *Institute for Environmental Studies, Vrije Universiteit, De Boelelaan 1115, 1081 HV Amsterdam*

Abstract

Remote sensing is an emerging technology with respect to water quality detection and monitoring. It must be made clear to end-users that the results of the technique are beneficial to them in their work. For this purpose it is necessary to provide the end-user with adequate water quality information from remote sensing at the right time, in the right format, at a competitive price (as compared to alternative methods). A methodology has been developed in The Netherlands, based on these criteria, applicable anywhere in the world. The methodology embedded in PC-based software is based on sound modeling of the water-atmosphere system. This makes it possible to derive accurate remote sensing algorithms for estimating water quality parameters for the types of water present. The case study in Friesland described here is a representative example of an applied inland water study.

1. Introduction

Water in all its phases is of prime concern to society. Remote sensing (from aircraft and satellites) and related environmental measurement methods offer unprecedented capabilities for synoptic measurement of many geophysical parameters. The scale of the measurements may vary from local (e.g. local water management authority level) to global (e.g. primary productivity of oceans).

Introduction of remote sensing of water quality is not hampered any more by lack of availability of sensors that are suited for the job. The real problem in getting an emerging accepted and implemented, lies in making it clear to end-users that application of the technique is beneficial in their work. Therefore it is necessary to provide the end-user with adequate water quality information from remote sensing at the right time in the right format, at a competitive price (as compared to alternative methods). For this purpose, a generic methodology has been developed in The Netherlands, applicable anywhere in the world. The following considerations were used as guidelines:

A. van Dijk and M. G. Bos (eds.), GIS and Remote Sensing Techniques in Land- and Water-management, 17–38.

1. None or few *in situ* measurements should be required.
2. Standardised operational methods and algorithms for processing of remote sensing data should be required to enhance reproducibility and to speed up the production process.
3. The development of analytical/deterministic algorithms should be a prerequisite to develop generally applicable methods with multisite, multisensor and multitemporal validity.
4. Algorithms must offer possibilities for sensitivity analyses and must be suitable to determine precision and errors of results.

The items currently detectable and measurable by remote sensing are biochemical and physical parameters that influence the underwater light field and the resultant physical characteristics of the underwater light field, and an indication of the significance and main use of these parameters (see table 1).

Table 1. The biochemical and physical parameters that influence the underwater lightfield and the resultant physical characteristics of the underwater lightfield:

Parameters that influence the underwater lightfield:

• *Chlorophyll a*
  • biomass, primary productivity
• *Cyanophycocyanin and cyanophycoerythrin*
  • specific cyanobacterial pigments
  • nuisance forming (toxic) algae;
  • eutrophication indicators
• *Seston dry weight*
  • sum of: algae, algal detritus, other organic and mineral matter
  • a measure of total suspended material
  • influences transparency
• *Aquatic humus (or dissolved organic matter)*
  • a tracer for waters of different origin
  • an essential component of the carbon contents of water

Parameters that are a function of the underwater lightfield:

• *Transparency (Secchi disk)*
  • swimming water standards
  • an often used easily measured but crude measure for water quality
• *Vertical attenuation coefficient (K_d)*
  • calculation of depth of euphotic zone
  • important for estimating establishment potential for macrophytes

The processing chain from light measured by a remote sensor to concentration maps of a water parameter is complex. By sound modeling of the water-atmosphere system it is possible to derive accurate remote sensing algorithms for estimating water quality parameters for the types of water. In the case study discussed these are river, lake and reservoir type water.

This approach based on physics enables:

1. The application of one algorithm to a time series of image data from the same sensor for the same region. This allows absolute intercomparison of the thematic water quality parameter maps.
2. All processing to be carried out on a PC; thus enabling transfer of the methodology at low costs.
3. A reduction in the amount and frequency of *in situ* and laboratory-based measurements. Only initial measurements are needed to establish the optical properties of the relevant waters in an area. Once the optical properties are established, repeat measurements are only required in case of calibration/validation activities or in the case of significant changes in the specific inherent optical properties (the concentrations may, however, vary significantly without requiring new measurements!).
4. Retrospective studies on satellite and airborne remote sensing data going back as far as Landsat MSS images from 1973. Because no *in situ* measurements are required at the time of the airborne or satellite images (as illustrated in the MH-DETEC (1997 and Dekker *et al.*, 1999b) study on coastal sediment concentration study in Kalimantan, where this methodology was previously applied), all historical data can be used.

The case studies discussed illustrate the validity of this approach. It also allows for more flexible remote sensing data gathering because there is no necessity for simultaneous *in situ* data collection. The acquisition and analysis of the field data is one of the bottlenecks for implementation of RS in countries less well equipped with a sophisticated sampling and laboratory infrastructure. Moreover, low cost, high quality remote sensing data comes within the reach of research and development groups, institutes and agencies with lesser purchasing power.

Figure 1 contains all required steps to establish algorithms, to perform a sensitivity analysis or to derive specifications for a remote sensing sensor. To establish algorithms only the "forward water" and the "inverse water" sections are relevant. To carry out a sensitivity analysis of the operational method or in order to derive specifications for a dedicated remote sensing instrument, it is necessary to go through all the steps from "forward water" via "forward atmosphere" to "inverse atmosphere" to "inverse water". Once the method is operational it is only necessary to run through the modules "inverse atmosphere" and "inverse water".

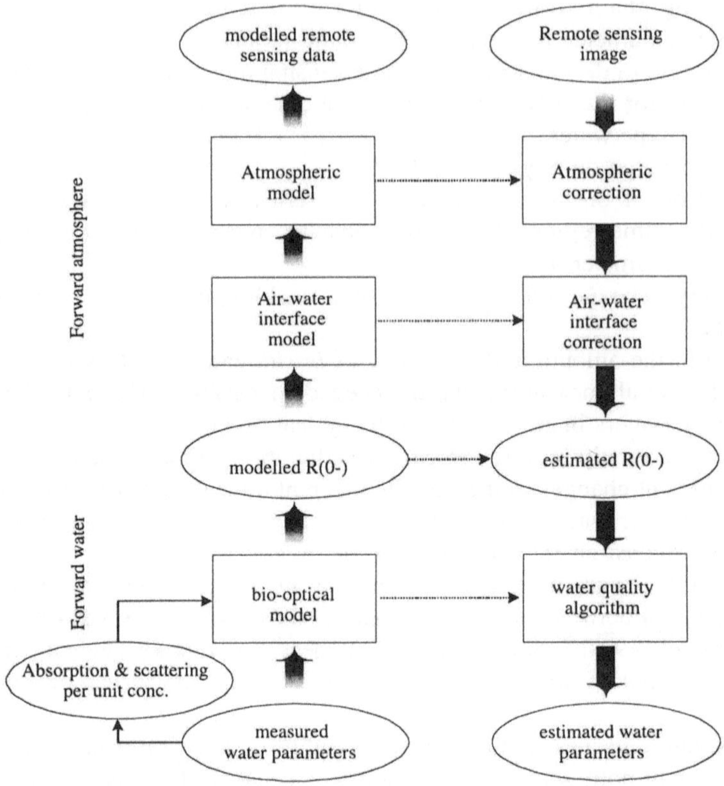

Figure 1. The forward and inverse model for remote sensing of water quality. In case of availability of measurements, the terms "modelled" or "estimated" may be replaced by measured.

2. Methodology for the extraction of water quality parameters from remotely sensed spectral data

2.1. Introduction

Dekker and Hoogenboom (1996) redefined the definitions originally determined by Morel and Gordon (1980) concerning possible approaches to estimate water quality parameters from remotely sensed spectral data. The three basis definitions were the *empirical method, the semi-empirical method and the analytical method.*

2.1.1. *The empirical method*

In the empirical approach statistical relationships are sought between measured spectral values (= "remote sensing image" section in Figure 1) and measured water parameters (see Figure 1). The limitation of such an approach is that spurious results may occur, because causal relationships between the parameters are not necessarily implied.

Results for the empirical models always need *in situ* data because the following parameters may change between different remote sensing missions:

above the air-water surface (the "forward" and "inverse atmosphere" sections in Figure 1)

- the total downwelling irradiance (solar elevation)
- the fraction of diffuse to direct solar irradiance
- the amount of specular reflection at the air-water interface
- the roughness of the water surface
- the height and composition of the atmospheric column between the sensor and the water surface leading to differences in path radiance

below the air-water interface (the "forward" and "inverse water" sections in Figure 1.)

- the radiance to irradiance conversion of the subsurface upwelling light signal
- the relation between the subsurface irradiance reflectance *R(0-)* and the inherent optical properties
- the relation between the inherent optical properties and the optical water quality parameters

2.1.2. *The semi-empirical method*

This approach may be used when the spectral characteristics of the parameters of interest are known. In Figure 1 this relates to the "forward water" section — but only the modules "measured water parameters" and "estimated or measured *R(0-)*" are used for the statistical analysis by focusing on well chosen spectral areas and appropriate wavebands or combinations of wavebands, which are used as correlates. Quantitatively, the *coefficients* from any such relationship only apply to the data from which they were derived. Each application must therefore be individually calibrated. This method is currently commonly used.

2.1.3. *The analytical or generic method*

(the "inverse atmosphere" and "inverse water" sections in Figure 1)

The inherent and apparent optical properties are used to model the reflectance and vice versa. The water constituents are expressed in their specific (per unit measure) absorption and backscatter coefficients. Subsequently a suite of analysis methods can be used to optimally retrieve the water constituents or parameters from the remotely sensed upwelling radiance or radiance reflectance signal. In Gordon and Morel (1983) a comprehensive discussion of the analytical models available for clear ocean waters through to turbid coastal waters till 1983 is given. Kirk (1983, 1994) extended the discussion to inland waters. Dekker (1993) developed approaches to analytical models for estimating chlorophyll (and cyanophycocyanin) from remote sensing data. Doerffer and Fischer (1994) present an approach for estimating concentrations of chlorophyll, suspended matter and Gelbstoff in Case II waters derived from satellite coastal zone colour scanner data with inverse modeling methods. Hoogenboom and Dekker (1998) successfully applied matrix inversion to a CASI image of a hypertrophic lake in The Netherlands.

3. The optical properties of natural waters

The relationship between the inherent optical properties, water quality parameters and the more measurable apparent optical properties of waters is displayed in section "forward water" in Figure 1. It is via their effects on the inherent properties that fluctuations in concentrations of the principal optical water quality parameters (water itself, dissolved aquatic humus, phytoplankton and non-living suspended particulate matter) affect subsurface reflectance and other measurements of underwater optical conditions.

For a proper discussion of remote sensing of water quality parameters it is necessary to define these basic optical properties of water and its constituents. The underwater light field is determined by the inherent optical properties (IOP) that are independent of the ambient light field (i.e., independent of changes in the angular distribution of radiant flux). These properties for light of a certain wavelength are specified by the absorption coefficient a (m^{-1}), the scattering coefficient b (m^{-1}), and the volume scattering function $\beta\ \sigma$ which describes the angular distribution of scattered flux resulting from the primary scattering process. For remote sensing purposes the backscattering coefficient b_b (m^{-1}) is relevant since this defines the amount of light scattered in an upward direction. The absorption coefficient of the medium as a whole, at a given wavelength, is equal to the sum of the individual absorption coefficients of the components present (Kirk, 1983). Assuming a linear relationship between concentration and absorption, the absorption coefficient due to any one component is proportional to the concentration of that component. Therefore:

$$a\ (total) = a\ (w) + a\ (ah) + a\ (ph) + a\ (t)$$

where $a\ (w)$ is the absorption by pure water, $a\ (ah)$ by CDOM (coloured dissolved organic matter: also referred to as aquatic humus, yellow matter, gilvin), $a(ph)$ by phytoplankton and $a\ (t)$ by tripton (non-chlorophyllous particles).

Scattering is the process by which photons change direction through interactions with matter and causes radiant energy to leave the water. More specifically it is backscattering that causes energy to leave the water (Kirk, 1991). Scattering is caused mainly by water $b(w)$, by phytoplankton $b(ph)$ and by tripton $b(t)$:

$$b = b\ (w) + b\ (ph) + b\ (t)$$

The spectral scattering of seston is a function of the size distribution and refractive index of the scattering particles. In productive and/or turbid coastal and inland waters the size distribution will be variable and complex.

The inherent optical properties have the function of being well-defined objective physical properties enabling the decomposition of a multitude of approaches and definitions into comparable properties. These inherent optical properties also give an insight into the implicit assumptions underlying many models used for remote sensing of water quality.

3.1. The nature of subsurface reflectance from rivers, lakes, coastal waters and oceans (the "forward water " section in Figure 1.)

Once the spectral absorption, scattering and backscattering properties have been determined for water samples it is possible to calculate specific absorption, scattering and backscattering coefficients, defined as the per unit spectral absorption and (back) scattering. Units are usually given in concentration per meter; e.g. useful measures are:

1. The amount of spectral absorption and (back)scattering caused by the equivalent algal biomass of 1 mg of chlorophyll *a*, referred to as the chlorophyll *a*-specific absorption and (back) scattering.
2. The amount of absorption and scattering caused by the non-algal part of the suspended matter (= tripton), referred to as the tripton-specific absorption and (back) scattering respectively.
3. The amount of absorption by aquatic humus (dissolved organic matter), referred to as the aquatic humus-specific absorption; here this quantity is related to the absorption at 440 nm.
4. Of course pure water has its own absorption and scattering properties.

In clear oceanic waters spectral reflectance is predominantly a function of (i) absorption by algal pigments and algal detritus and low concentrations of aquatic humus at short wavelengths and by pure water at longer wavelengths, (ii) scattering by water molecules at short wavelengths, Raman scattering at intermediate wavelengths and algal cell material, (iii) the solar stimulated fluorescence of algal pigments at longer optical wavelengths.

In turbid coastal waters, and in almost all inland waters, these effects also occur but the analysis becomes more complicated due to the importance of two additional components. First, backscattering from particles is the dominant scattering factor. This can be up to 1000 times the backscattering of the clearest oceanic waters. Second, reflectance at short wavelengths is reduced due to absorption by high concentrations of dissolved humic compounds, algal pigments and particulate matter.

The IOP are physically related to the subsurface irradiance reflectance *R(0-)* which is a key parameter linking the IOP to remote sensing or *in situ* (ir)radiance measurements. *R(0-)* is only slightly dependent on solar elevation, atmospheric or water surface conditions and is therefore called a quasi-inherent property. Algorithms developed on the basis of *R(0-)*, irrespective of the mode of calculation or measurement, have multitemporal validity.

The relation between *R(0-)* and the inherent optical properties (IOP) for ocean and inland waters systems was investigated by (Gordon *et al.,* 1975, Morel and Prieur, 1977; Whitlock *et al.,* 1981 ; Kirk, 1991; Dekker, 1993; Dekker *et al.,* 1997). Dekker *et al.,* 1997, found that the following linear backscattering albedo model was the most appropriate model for coastal and inland waters:

$$R(0-) = r_1 \frac{b_b i\,(\lambda)}{a_i\,(\lambda) + b_b i\,(\lambda)}$$

where r_I is a coefficient for the anisotropy of the lightfield; b_{bi} is the backscattering coefficient for parameter *i*; and a_I is the absorption coefficient for parameter *i.*

The IOP of all four optically active components (i) are included in the model: phytoplankton, tripton, aquatic humus and water.

3.2. The "water forward" model: from SIOP to *R(0-)*

The bio-optical model is used to simulate the effect of changes in an optical water quality parameter such as e.g. chlorophyll ((CHL) concentration on *R(0-).* Figure 2. presents a run of the model for estimating the effect of changing chlorophyll concentrations (from 20 to 200 μg^{-1} in steps of 20 $\mu g\ l^{-1}$) in a shallow, eutrophic lake system with high aquatic humus concentrations. As can be seen in the table at the top of Figure 2. the resultant increase in seston dry weight is calculated as well as the Secchi disk transparency and the vertical attenuation coefficient for PAR. As input appropriate values for each of the specific inherent optical properties were determined, i.e. the inherent optical properties per unit water quality parameter: e.g. the specific inherent absorption by chlorophyll α, a^*_p, is the amount of absorption caused by 1 $\mu g\ l^{-1}$ chlorophyll-α.

$$a = a^*_{ph}\ CHL + a^*_t\ C_t + a^*_{cdom}\ (440) + a_w$$

$$b_b = b^*_{b_{ph}} CHL + b^*_{b_t}\ C_t + b_{b_w} = b^*_{b_s}\ \{CHL + C_t\} + b_{b_w}$$

Specific IOP are denoted by an asterisk as superscript. Values for the IOP in this model are:

- a_w : The temperature dependent absorption by pure water from Buiteveld et al. (1994).
- b_{bw} : The backscattering of pure water from Buiteveld et al. (1994).
- a^*_{ph} : Specific inherent absorption by chlorophyll (derived from Dekker (1993).
- a^*_t : Specific inherent absorption by tripton from Dekker (1993).
- b^*_{bs} : Specific inherent backscattering of seston (i.e. the sum of tripton and phytoplankton), from Dekker (1993).
- $a^*_{cdom}(440)$: The specific absorption coefficient of aquatic humus at 440 nm is calculated by assuming an exponential slope for aquatic humus absorption:

$$a^*_{cdom}(\lambda) = e^{S(\lambda - 440)}$$

values of S range from 0.01 to 0.02 nm^{-1} (Dekker, 1993, Kirk, 1994) in most natural waters. In this case a value of 0.15 was used.

run	1	2	3	4	5	6	7	8	9	10
DW	21.4	22.8	24.2	25.6	27.0	28.4	29.8	31.2	32.6	34.0
CHL	20	40	60	80	100	120	140	160	180	200
a_cdom	3.0	3.0	3.0	3.0	3.0	3.0	3.0	3.0	3.0	3.0
Ct	20.0	20.0	20.0	20.0	20.0	20.0	20.0	20.0	20.0	20.0
Kd	2.2	2.4	2.7	2.9	3.1	3.4	3.6	3.8	3.9	4.1
SD	0.48	0.45	0.42	0.40	0.38	0.36	0.34	0.33	0.31	0.30

Main menu

Output

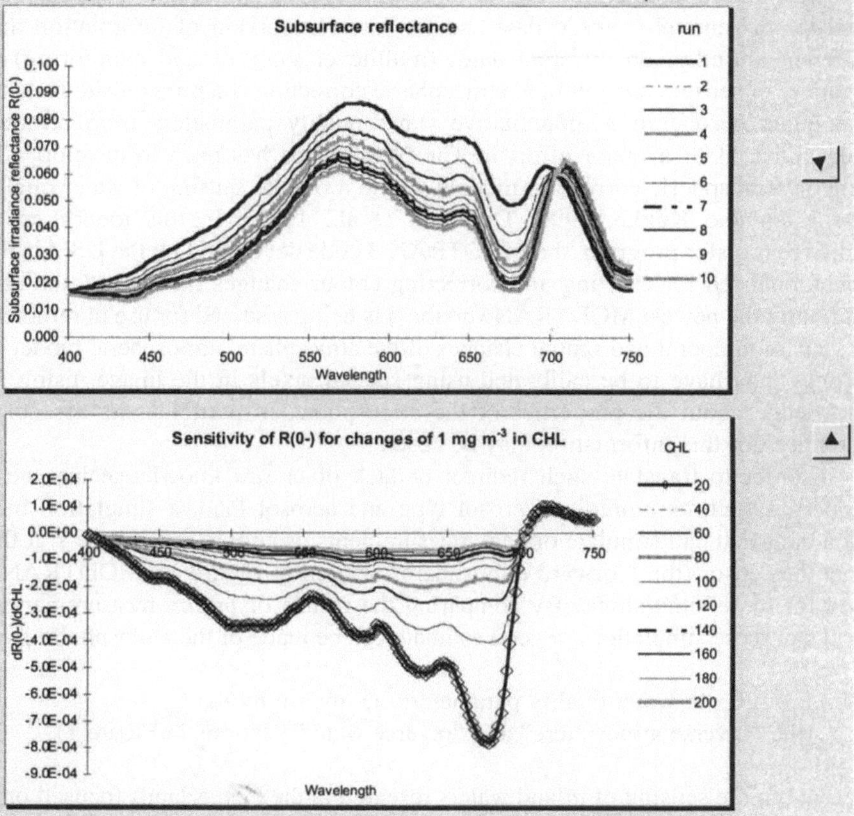

Figure 2. Run of the bio-optical model for estimating the effect of changing chlorophyll concentrations (from 20 to 200 μg^{-1} in steps of 20 μg l^{-1}) on reflectance in a shallow, eutrophic shallow lake system with high aquatic humus concentrations. The upper graph shows the modeled subsurface irradiance reflectance curves; the lower graph shows the change in reflectance, caused by adding 1 μg l^{-1} chlorophyll, at a concentration of 20, 40 ...200 μg l^{-1} depicted as d*R(0-)*/dCHL. It is clear that over the entire wavelength range studied the sensitivity of *R(0-)* to increasing chlorophyll levels becomes less as the concentration increases. Explanation of parameters in table: DW= seston dry weight in mg l^{-1}; CHL= concentration of chlorophyll *a* and phaeophytine in μg l^{-1}; a_{cdom} is the absorption by dissolved organic matter at 440 nm; Ct = the concentration of tripton — the non-algal part of the DW; Kd = the vertical attenuation coefficient for PAR; SD is the Secchi disk transparency in m.

3.3. The "forward atmosphere" model: from *R(0-)* to L_{rs}

The effect of atmospheric scattering and absorption on the total remotely sensed signal can be substantial, especially in the case of water bodies, which are relatively dark. Accurate and reliable correction for those effects is of great importance for quantitative analysis of remotely sensed data and for intercomparison of information acquired at different altitudes, on different dates (multilevel analyses and monitoring) and from a variety of sensors. Operational atmospheric correction is a prerequisite for an adequate and quick derivation of quantitative water quality parameters from remote sensing data. Since 1993 a major effort in The Netherlands has been to develop a PC-based generic atmospheric correction module within a remote sensing of water quality toolkit (De Haan and Kokke, 1996; De Haan et al., 1997). In this toolkit an advanced radiative transfer program, the MODTRAN 3 code developed by the US Air Force, was operationalised for charting and correcting colour changes induced by the atmosphere (currently the newest MODTRAN version 4 is being assessed for use in remote sensing). In view of temporal and spatial changes of the atmosphere, atmospheric model parameter settings may have to be calibrated using special pixels in the image, using additional knowledge about the properties of the atmosphere, in particular its aerosol load and humidity, for this information may be absent.

In order to translate such indirect or lack of *in situ* knowledge into atmospheric properties such as humidity, aerosol type and aerosol load, a simulation module has been built that can simulate optical measurements or known reflectance's at the surface or at the sensor (the "forward atmosphere" section in Figure 1). MODTRAN 3 is also used for these simulations. By comparing the results of *in situ* measurements with the results of these simulations, a good estimate can be made of the atmospheric properties.

3.4. From L_{rs} to water quality parameter: the inverse model (the "inverse atmosphere" and "inverse water" sections in Figure 1)

Recent remote sensing of inland waters research in the Netherlands focused on developing multitemporally valid algorithms based on the analytical model. This requires knowledge of the underlying inherent optical properties of the water and its constituents. Making use of the 1990 remote sensing, *in situ* and laboratory data collected during the 1990 CAESAR and CASI airborne remote sensing flights, Dekker (1993) developed an analytical model for estimating chlorophyll-a and cyanophycocyanin from high spectral resolution data, calibrated to *R(0-)*. This approach requires the estimation of specific absorption and backscattering coefficients for both algal and non-algal particulate matter. These were determined from spectrophotometric measurements of apparent absorption and scattering. Chlorophyll *a* could be derived from the airborne estimates of *R(0-)* with an accuracy of 9.5 mg m^{-3} and cyanophycocyanin with an accuracy of approximately 20 mg m^{-3}. These analytically derived algorithms are based on the subsurface irradiance reflectance and were subsequently applied successfully to several airborne remote sensing campaigns in the same region (Van Kootwijk *et al.* 1996 and Dekker *et al.*, 1999a), thus illustrating the multitemporal validity of analytical algorithms based on the *R(0-)*. In Van Kootwijk *et al.* (1996) the same algorithms were applied to another lake area in the Netherlands: i.e. Frisian Lakes in stead of Vecht Lakes. The form of the algorithm's remained the same, the coefficients, however, had to be calculated making use of the *in situ* data. Thus these analytical algorithms are site-specific.

4. Case Studies

Two case studies are presented which illustrate the validity of the approach and which clearly show the previously mentioned advantages. It is also allows for more flexible remote sensing data gathering because there is no necessity for simultaneous *in situ* data collection — this being one of the bottlenecks for implementation of RS in countries less well equipped with a sophisticated sampling and laboratory infrastructure. Moreover, high quality remote sensing comes within the reach of smaller research and development groups, institutes and agencies.

The two examples are:

1) two flights using a CASI imaging spectrometer over a lake/wetland area in Friesland in the northwest of the Netherlands, illustrating the high level of sophistication available for remote sensing of wetlands (Plate 1)
2) the analysis of 3 Landsat TM images (1995) and a SPOT image (1995) for determining the natural conditions of flow and sedimentary transport in the Frisian Lakes (Plate 2).

4.1. Case 1: Frisian wetlands/lakes by airborne imaging spectrometry (after Dekker *et al.*, 1999a)

The wetland/lake complex "Alde Faenen" lies in the northeast of The Netherlands with an overall size of 6 by 8 km (see Plate 1). It is a complex of partly partitioned lakes. Canals connect these waters to the regional watershed. These waters have an important ecological function within this large agricultural area. It was originally a low peat moor area. Over-extraction of turf in past centuries caused the formation of the lakes. Up till 1989 this wetland area became increasingly eutrophied due to inlet of nutrient rich water from Lake IJsselmeer in summer and agricultural runoff from the surrounding area in the winter. Since 1989 restoration measures have been taken, whereby different methods were used, in order to study their effects. Some of these measures were biomanipulation by fishing out whitefish, and isolation and dredging to developing wetlands by inundating former agricultural fields. The net result of these measures, as determined from CASI data in the late summer of 1995 and 1997 are shown in Plate 1.

In 1995 an intensive measurement campaign of over 115 Dutch water was carried out in order to get a grip on the variation in optical water quality in The Netherlands and to establish a database for further use in remote sensing of water quality. The results are presented in Rijkeboer *et al.*, (1997, 1998). 18 lakes and canals in the province of Friesland were measured. The range in optical water quality parameters was large as is illustrated in Table 2.

State-of-the-art in airborne remote sensing of inland water quality

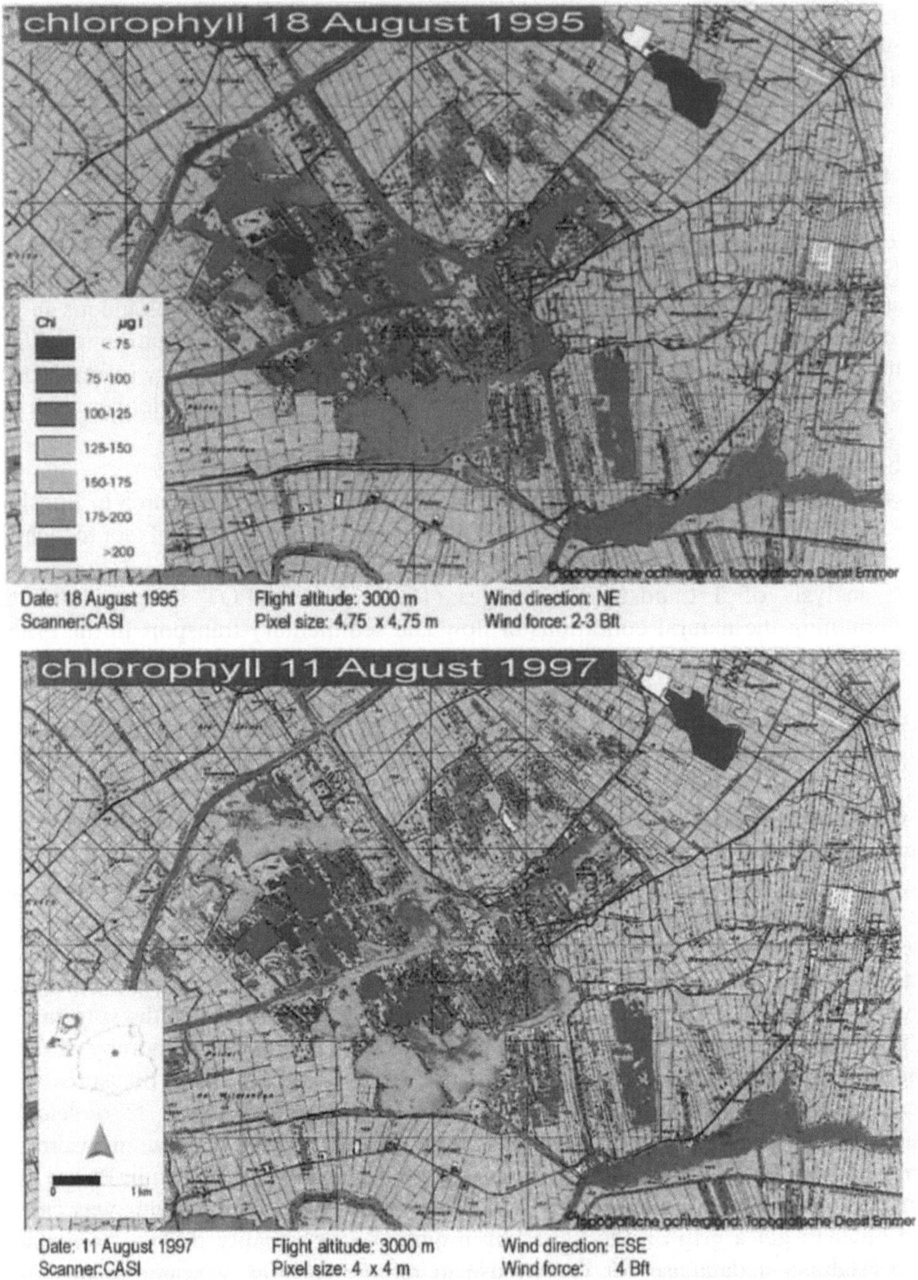

Plate 1. This map shows the results for assessing the chlorophyll concentration from two CASI flights carried out in the late summer of 1995 and 1997. The same algorithms were applied after prior atmospheric correction. Only one legend is required. In 1997 no *in situ* data were required. Deterioration in water quality is visible from 1995 to 1997, based upon overall increasing levels of chlorophyll concentration. For the local water management authorities such maps are important because they are incapable of monitoring each of these lakes and canals (Dekker *et al.*, 1999a).

Plate 2. A Landsat TM satellite image from 12 August 1995 is shown that was processed using the analytical algorithm in the text. The same algorithm was applied both times after correcting the images for atmospheric and air/water interface effects. The model was initiated on measurements performed in 1995. No *in situ* measurements were available on the days of the satellite image recording (from Dekker *et al.*, in press a).

Table 2. Values of optical water quality measurements measured in Friesland in August 1999

	lowest		highest
CHL ($\mu g\ l^{-1}$)	2	-	343
CPC ($\mu g\ l^{-1}$)	0	-	1000
DW ($mg\ l^{-1}$)	2	-	53
SD (m)	0.2	-	1.1
$a_{(ah)440}$ (m^{-1})	0.1	-	16

* note: the lowest values of concentration parameters do not coincide with the highest values of transparency because the low values occurred in shallow lakes with a depth of 60 cm's.

The standard analytical (as opposed to empirical!) algorithm used (Figure 3) is a ratio of two spectral bands that are located exactly where (see Figure 2) the effects of chlorophyll a absorption on the subsurface irradiance reflectance are highest: at 676 nm and at 706 nm. The CASI flown in these flights had 15 spectral bands of approximately 10 nm, with two bands centered at these wavelengths. The same algorithms were used in both years. Overall, despite the measures taken to improve the lakes, deterioration in water quality is visible: several lakes that were in the lower chlorophyll regions in 1995 contain significantly higher concentrations of chlorophyll in 1997.

From these measurements using the methodology explained above and presented in Figure 1, relationships were determined between the optical water quality parameters and the inherent optical properties, thereby enabling calculation of the specific inherent optical properties.

From the SIOP's and the known concentrations, the *R(0-)* was simulated and, in turn, these were calibrated with the *in situ* measured spectroradiometric *R(0-)*. The *in situ R(0-)* measurements were not performed on the remote sensing day, but in the week before the flight. Because synchronous atmospheric composition data (based on field measured data) was not available, an iterative procedure was followed to determine the atmospheric correction parameters at the time of the remote sensing flight. First the standard available data on meteorological circumstances was collected (e.g. humidity, temperature, air mass origin, wind direction and speed). From the radiance's measured at the CASI sensor, using the Toolkit software for atmospheric correction and simulation, an atmospheric composition was iteratively approximated that was in overall accordance with the expected reflectance's based on both simulated and measured *R(0-)*.

Algorithms derived in earlier work from another similar wetland area in The Netherlands (Dekker, 1993), the Loosdrecht Lakes were adapted (through bio-optical modelling) to conform to the higher concentrations of aquatic humus and the larger range of optical water parameters encountered here. Because the predicted relationship was well described by an exponential function, this function was used for inverting the model. Figure 3 shows this exponential function together with the fit to the actual measured *in situ* data. (Note that the *in situ* measured data was not measured at the time of the overflight — thus the results are merely an indication of the validity; also note that the relationship in Figure 3 is a test of the modelled relationship).

In 1997 no *in situ* measurements were made during the airborne remote sensing flight. The same procedure for the 1995 flight was followed. Lakes with known low concentrations were used to fine-tune the atmospheric correction. Once the *R(0-)* had been calculated, the same algorithms from 1995 were applied. Plate 1 shows the 1995 and 1997 thematic water quality maps. The local water management authorities did sample some parameters in the weeks surrounding the date of the flight, and these were in good accordance with the estimated parameters from the imaging spectrometry flight. The success illustrates the validity of this approach.

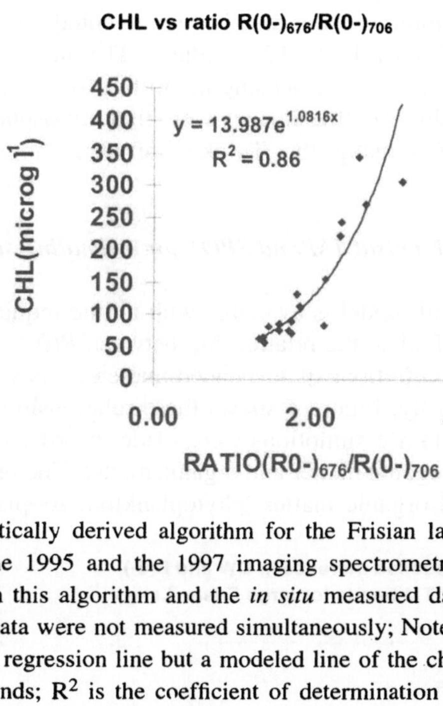

Figure 3. The analytically derived algorithm for the Frisian lakes derived in August 1995 and applied to both the 1995 and the 1997 imaging spectrometry campaign flight data, and the correlation between this algorithm and the *in situ* measured data for August 1995 (note 1: *in situ* data and flight data were not measured simultaneously; Note 2: the black curvilinear line in this figure is NOT a regression line but a modeled line of the chlorophyll to reflectance ratio of the used spectral bands; R^2 is the coefficient of determination for the relationship between the modeled line and the *in situ* data.).

4.2. Case 2: Frisian lakes by satellite imagery (after Dekker *et al.*, in press a & b)

Using the same methodology as above an algorithm was developed for assessing the TSM concentration in all of the southwest Frisian lakes using satellite data from Landsat and SPOT (see Vos et al, 1998, for an extensive discussion and for colour illustrations).

Using the bio-optical model it was possible to model the *R(0-)* in full spectral resolution (typically at 1 or 2 nm resolution over the range of 380 to 800 nm or more). Next the *R(0-)* spectra were recalculated to the spectral bands of the instrument applied, in this study the Landsat TM and SPOT sensors. It is important to realize that the Landsat TM and SPOT sensors average the reflectance values over the entire width of the spectral band; all spectral information within one band is averaged as well (these bands are typically 60 to 80nm wide in the visible wavelength domain). In the simulations average values were used for concentrations of chlorophyll and dissolved organic matter; next these spectra can then be plotted as a function of increasing concentrations of DW which automatically gives the analytical relationship with reflectance in the spectral bands of the sensor. Because this is a model simulation all possible concentrations can be modeled: thus the often encountered problem of a non-uniformly distributed *in situ* data set which had to be used for estimating water quality parameters from remote sensing data is circumvented (as occurred in the DHV (1988) and Roeters and Buiteveld (1993) studies). The next step is to invert this relationship. Which leads to a remote sensing algorithm for the parameter of interest. Of course, the remote sensing satellite images were first atmospherically and air/water interface corrected to *R(0-)* using the Toolkit software (an extended version of MODTRAN 3).

4.2.1. *The Algorithms for Landsat TM and SPOT for estimating suspended matter*

Once the forward bio-optical model is available with all the required input parameters, it is relatively simple to calculate the relationship between *R(0-)* (in each of the sensor bands) and increasing concentrations of suspended matter, expressed in terms of seston dry weight (DW) in this study. Figure 4 shows these relationships for the four optical bands of Landsat TM. Certain assumptions were made in order to obtain a consistent data set. DW consists of inorganic matter and organic matter. The organic matter consists of phytoplankton and dead organic matter (phytoplankton, zooplankton, macrophytes, peat detritus etc.).

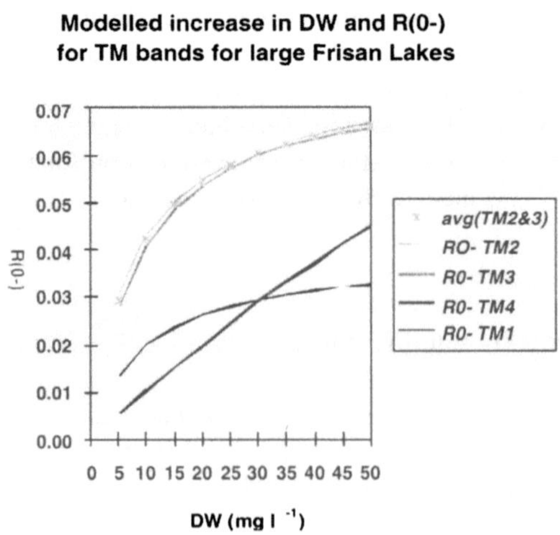

Figure 4. The analytical relationship between *R(0-)* in Landsat TM Bands 1,2,3 and 4 and TSM for an average set of Frisian large lakes in the south-west of the province; based on bio-optical modelling (adapted from Dekker *et al.*, in press b).

Because the sensors used cannot discriminate within their spectral bands any pigments from suspended matter it was decided to assume average values for the ratio of organic to inorganic material and of the amount of DW that is composed of phytoplankton.

The relationship of Landsat TM Band 1 and DW concentrations is the least suited for estimating DW because the combined effect of all the absorbing parameters causes the relationship to reach low asymptotic levels after about 15 mg l^{-1}. Also the absolute range between 1.2% and a maximum 3% *R(0-)* is low. A further reason not to use this band is that the atmospheric contribution and the reflection of diffuse sky irradiance at the water surface is maximal in this spectral band. The spectral bands 2 and 3 of TM (more or less the same as B1 and B2 for SPOT) show an almost equivalent behavior with increasing DW concentrations. They start at 3% *R(0-)* for 5 mg l^{-1} and rapidly increase to 6% at 30 mg l^{-1} after which saturation occurs. Judging from the graph the relationship of band TM4 with increasing DW levels is the most suited for inversion: it is an almost linear increasing function of increasing *R(0-)* with increasing DW. Because there are no effects of variable absorbing parameters anymore; it is solely a function of increased scattering caused by increased levels of particulate material. Nevertheless TM4 is not selected for use because the sensor is relatively less sensitive in this domain (i.e. a small number of DN's encompasses the full range shown here) and the adjacency effect probably plays a significant role due to the high contrast of vegetation reflectance on land and the low water targets. The adjacency effect was studied using the Toolkit based MODTRAN module. Unfortunately it was only possible to determine the adjacency effect of adjacent pixels whereas it would be most interesting to estimate the effect of high background pixels on all pixels of a lake.

Thus, Landsat TM Bands 2 and 3 are most suitable for estimating DW for the Frisian large lakes. In order to decrease the sensitivity of an algorithm to variations in tripton and aquatic humus absorption (having more effect in band 2) and to variations in chlorophyll absorption (more prominent in Band 3), it was decided to take the average of Band 2 and Band 3 for development of an algorithm. Figure 5 shows the algorithm for estimating DW from Landsat (for SPOT a similar relationship was found).

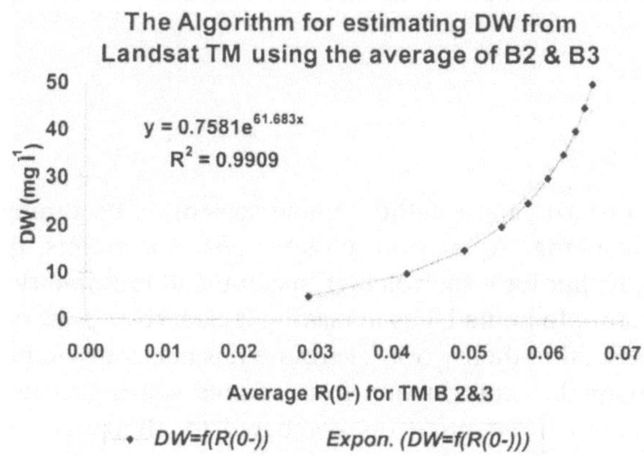

Figure 5. The analytical relationship between DW and the average of Landsat Bands 2 and 3 approximated by an exponential relationship. The exponential relationship was used as the algorithm for estimating DW from the Landsat satellite images. Note: the dotted curvilinear line is NOT a regression line but a modelled line of the DW to TMB2 and B3 relationship.

These relationships were so close to a power function (r^2 = 0.99) that for both sensors, for reasons of ease of digital image processing, the power function was used as the algorithm for estimating DW. It must be noted here that the relationships remain analytical and are not based on any type of regression analysis; the analytical relationship is merely approximated by a power function.

The methodology developed and presented here for processing and analysing the TSM maps based on Landsat TM and SPOT allows multi-temporal, multi-site and multi-instrument comparison. This is considered to be a breakthrough in the area of remote sensing applications. Now satellite data can become an independent measurement tool for water management authorities. Hitherto it was always necessary for water management authorities to organise a massive *in situ* measurement campaign exactly coinciding with a satellite overpass, in order to obtain sufficient data for empirical analysis relating TSM measured *in situ* to (often raw DN) data of a satellite sensor. This was often hampered by logistical difficulties and, was, most of the time, an unsuccessful approach. Based on the algorithms developed here it is possible for Waterschap Friesland to acquire Landsat TM and SPOT images and to have them processed at relatively low-costs to TSM maps. The models for forward and inverse modelling of these lakes may become more sophisticated in future, but this is now a matter of choice and not necessity.

4.2.2. *Suspended matter maps of the Frisian Lakes*

The results of an analysis of the suspended matter maps derived from remote sensing satellite images of Landsat TM and SPOT are shown in Table 3. These suspended values reveal the range and average concentration of suspended matter for the Southern Frisian Lakes for 24 May, 11 July, 2 August (Fluessen not available) and 12 August 1995 (Plate 2).

The maps shown in Dekker *et al.*, in press b, Plate 2 and the values of TSM in Table 3 clearly demonstrate how valuable remote sensing can be to understand a water system. There is no *in situ* measurement campaign that can deliver such spatially exhaustive data for a lake.

5. Conclusions

The technology of airborne and satellite remote sensing is becoming well-established, it is also rapidly advancing. Apart from possible cost constraints, the real problem in getting an emerging technology such as high spectral and radiometric resolution remote sensing accepted and implemented, lies in making it clear to end-users that application of the technique is beneficial in their work. Therefore it is necessary to provide the end-user with adequate water quality information from remote sensing at the right time in the right format, at a competitive price (as compared to alternative methods). For this purpose, a generic methodology has been developed in The Netherlands, applicable anywhere in the world. The following considerations were used as guidelines:

Table 3. Statistics of remote sensing TSM maps for the Frisian Lakes (after Dekker *et al.*, in press a).

	Fluessen	Sneeker-meer	Groote Brekken	Koevorder-meer	Tjeuke-meer	Sloter-meer
24-May-95						
Average	21.0	8.7	7.9	7.0	13.4	61.6
st. deviation	5.6	2.9	2.0	1.7	4.0	19.0
minimum	9	3	3	3	5	21
maximum	59	24	17	19	36	154
11-Jul-95						
Average	13.3	11.4	7.5	9.6	19.7	19.5
st. deviation	2.6	3.1	1.6	2.1	4.8	3.3
minimum	6	4	4	5	10	8
maximum	29	39	21	21	48	35
02-Aug-95						
Average	x	25.6	25.3	15.8	92.3	73.9
st. deviation	x	8.7	8.5	4.5	32.3	13.9
minimum	x	10	11	8	30	11
maximum	x	411	111	79	245	157
12-Aug-95						
Average	19.9	13.3	26.4	22.9	39.4	89.3
st. deviation	4.1	2.9	5.4	4.5	12.3	30.1
minimum	9	6	14	14	13	18
maximum	44	31	65	65	119	255

1. In operational remote sensing none or few *in situ* measurements should be required, as concurrent sampling and analysis is expensive and labour intensive.
2. Standardised operational methods and algorithms for processing of remote sensing data are required to enhance reproducibility and to speed up the production process. Remote sensing derived information should be able within the same timeframe as laboratory results become available.
3. The development of analytical/deterministic algorithms is a prerequisite to develop generally applicable methods with multisite, multisensor and multitemporal validity. Empirical-based approaches always need a large amount of *in situ* data.
4. Algorithms must offer possibilities for sensitivity analyses and must be suitable to determine precision and errors of results. Thus a end-user knows whether or not a particular remote sensing application meets their requirements.

Two case studies are presented where the method has been applied with success: 1) two flights using a CASI imaging spectrometer over a lake/wetland area in the northwest of the Netherlands as an example of the high level of sophistication available for remote sensing of wetlands and 2) the analysis of Landsat TM (1995) and SPOT images for determining the TSM values and ranges in Frisian lakes.

The methodology presented here is available in prototype software, which upon completion may provide low-cost capabilities all over the world for making more meaningful use of remote sensing data. Furthermore the advent of Internet/WWW capabilities worldwide, will make it possible to apply such methods anywhere more easily, with less intensive (= less expensive) local support by experts, but with scientific and technical support coming from those experts at their home base. Distributed processing and distance-training will be keywords in the coming decade.

5. Recommendations

It is necessary to provide the end-user with adequate water quality information from remote sensing at the right time in the right format, at a competitive price (as compared to alternative methods). Providing cost benefit analyses for remote sensing derived information is required; however, this may not be easy to perform as the type of information remote sensing may offer is of another nature than traditional water quality assessment methods. Remote sensing gives a complete spatial image of an area, with a limited number of parameters. The information on water quality is only available from within the penetration depth of light in the water column (and within that range exponentially decreasing with depth). The frequency of availability of remote sensing data is an important issue. Frequency is more easily obtained from satellites at lower costs than associated with airborne remote sensing: but a boundary condition is that the satellites have a sufficient temporal coverage to provide an adequate amount of sufficiently cloud free images per required monitoring period. Airborne remote sensing offers greater flexibility, but at higher cost per spatial unit. In general, airborne platforms offer more advanced systems being able to estimate more parameters. Thus a cost-benefit analysis is subjective to the spatial, temporal and accuracy requirements of the end-user.

To involve the end-users it is recommended to carry out more demonstration/test case studies at various geographic locations in the world. In The Netherlands there is currently available a suite of PC-based WINDOWS software tools (albeit in a prototype form) associated with an overall analytical approach to estimating the above mentioned set of water quality parameters from remote sensing data from any type of optical sensor, in any type of water, at any location in the world. It is continuously being improved and currently a dialogue is open how to further operationalize and possibly commercialise this software.

6. References

Buiteveld, H., J.H.M. Hakvoort and M. Donze. (1994) Optical properties of pure water. In: J.S. Jaffe, editor. *Ocean Optics XII,* 6-1994, SPIE, Bellingham, Washington, USA, p. 174-183.

De Haan, J.F. and J.M.M. Kokke, (1996) *Remote sensing algorithm development Toolkit I: Operationalization of atmospheric correction methods for tidal and inland waters,* BCRS 96-16, BCRS, Delft, 91 p.

De Haan, J.F., J.M.M. Kokke, H.J. Hoogenboom and A.G. Dekker, (1997) An integrated toolbox for processing and analysis of remote sensing data of inland and coastal waters-atmospheric correction. *Fourth International Conference: Remote Sensing for Marine and Coastal Environments,* 3-1997, ERIM, Michigan, USA, p. 1-215-1-222.

Dekker, A.G., (1993) *Detection of optical water quality parameters for eutrophic waters by high resolution remote sensing.* PhD. Thesis, Vrije Universiteit, Amsterdam, The Netherlands; 240 p.

Dekker, A.G. and H.J. Hoogenboom. (1996) Predictive modelling of AVIRIS performance over inland waters. In: R.O. Green, editor. *Sixth Annual JPL Airborne Earth Science Workshop,* —32767, JPL Laboratory, Pasadena, California, USA.,

Dekker, A.G. and H.J. Hoogenboom, (1996) *Operational tools for remote sensing of water quality: a prototype Toolkit,* BCRS report 96-18, BCRS, Delft, The Netherlands, 66 p.

Dekker A.G., Moen J.P., Kootwijk E.J., vna Rossum G., Hoogenboom H.J., *et al.*, 1999a. The water quality of some Frisian lakes measured by airborne remote sensing (August 1997). *Report NRSP-2 99-22*, Netherlands Remote Sensing Board, Programme Bureau, Rijkswaterstaat Survey Department, Delft, The Netherlands.

Dekker, A.G., S.W.M. Peters, M. Rijkeboer, and H. Berghuis. (1999b) Analytical processing of multitemporal SPOT and Landsat images for estuarine management in Kalimantan Indonesia. In: G.J.A. Nieuwenhuis, R. Vaughan, and M. Molenaar, editors. *18th EARSeL Symposium on operational remote sensing for sustainable development,* A.A.Balkema, Rotterdam,The Netherlands, p. 315-324.

Dekker, A.G., H.J. Hoogenboom, L.M. Goddijn and T.J.M. Malthus, (1997) The relationship between inherent optical properties and reflectance spectra in turbid inland waters. *Remote Sens. Reviews,* Vol. 15, p. 59-74.

Dekker, A.G., H.J. Hoogenboom, M. Rijkeboer and H. Buiteveld. (1997) Estimation of water quality parameters of inland waters with the use of a toolkit for processing of remote sensing data. *Remote Sensing for Marine and Coastal Environments: Technology and Applications,* 3 -1997, ERIM, Michigan, USA, p. 1-269-1-275.

Dekker A.G., Vos R.J., Peters S.W.M., *et al.*, *in press a.* Comparison of Remote Sensing Data, Model Results and In-Situ Data for the Southern Frisian Lakes. *The Science of the Total Environment.*

Dekker A.G., Vos R.J., Peters S.W.M., *et al.*, *in press b.* Analytical algorithms for lake water TSM estimation for retrospective analysis of TM and SPOT sensor data. *International Journal of Remote Sensing.*

DHV raadgevend ingenieursbureau B.V., (1988) *De toepassingsmogelijkheden van remote sensing satellietopnamen voor het waterkwaliteitsbeheer van de Friese meren,* BCRS Report, Delft, The Netherlands.

Doerffer, R. and J. Fischer, Concentrations of chlorophyll, suspended matter and gelbstoff in case II waters derived from satellite coastal zone color scanner with inverse modeling methods. *Journal of Geophysical Research,* (1994) Vol. 99, No. c4, p. 7457-7466.

Gordon, H.R., O.B. Brown, M.M. Jacobs, (1975) Computed relationships between the inherent and apparent optical properties of a flat homogeneous ocean. *Appl. Opt.,* Vol. 14, No. 2, p. 417-427.

Gordon, H.R. and A. Morel, (1983) *Remote assessment of ocean color for interpretation of satellite visible imagery: a review,* Springer-Verlag, New York, 114 p.

Hoogenboom, H.J., A.G. Dekker and J.F. De Haan, (1998) Retrieval of chlorophyll and suspended matter in inland waters from CASI data by matrix inversion. *Can. J. Remote Sensing,* Vol24, no 2: p 144-152.

Kirk, J.T.O., (1983) *Light and photosynthesis in aquatic ecosystems,* Cambridge University Press.,

Kirk, J.T.O., (1991) Volume scattering function, average cosines, and the underwater light field. *Limnol. Oceanogr.,* Vol. 36, No. 3, p. 455-467.

Kirk, J.T.O., (1994) *Light & photsynthesis in aquatic ecosystems,* 2 ed., 509 p.

Kootwijk, E.J., A.G. Dekker, H.J. Hoogenboom, J.P. Moen, G.A. van Rossum and B.J. Schoenmakers, (1996) *Remote sensing for managing inland waters (in Dutch),* BCRS report 95-29, BCRS, Delft, The Netherlands.

Moen, J.P., J.D. van Setten, E.J. van Kootwijk, A.G. Dekker, H.J. Hoogenboom, G.A. van Berkum, T.H.L. Claassen, B. van der Veer, (1997) *De kwaliteit van Nederlandse binnenwateren gemeten met vliegtuig remote sensing (1995),* NRSP-2 96-27, BCRS-PB, Survey Department, Rijkswaterstaat,The Netherlands, Delft, The Netherlands, 61 p.

MH-Detec, (1997) *The study to the access channel to the port of Banjarmasin:the natural conditions of flow and sediment transport in the estuary and delta area,* 97.016.X1./FR1, MH-NL BV, Hendrik Ido Ambacht, The Netherlands, 119 p.

Morel, A. and L. Prieur, (1977) Analysis of variations in ocean colour. *Limnol. Oceanogr.,* Vol. 22, No. 4, p. 709-722.

Morel, A. and H.R. Gordon, (1980) Report of the working group on water color. *Boundary-Layer Meteorology,* Vol. 18, p. 343-355.

Roeters, P.B. and H. Buiteveld, (1993) *Use of satellite imagery assessment of the eutrophication in the Frisian Lakes,* RIZA, 90 p.

Rijkeboer, M., A.G. Dekker and J.M.M. Kokke, (1997) *Kleurmetingen van Nederlandse oppervlaktewateren,* NRSP-2 96-14, BCRS-PB, Survey Department, Rijkswaterstaat, Delft, The Netherlands, 57 p.

Rijkeboer, M., H.J. Hoogenboom and A.G. Dekker, (1997) *Realisatie spectrale bibliotheek van Nederlandse wateren,* W-97/07, 48 p.

Rijkeboer, M., A.G. Dekker and H.J. Gons, (1998) Subsurface irradiance reflectance spectra of inland waters differing in morphometry and hydrology. *Aquat. Ecol.,* Vol. 31, p. 313-323.

Vos, R.J., A.G. Dekker, S.W.M. Peters, G. van Rossum and L.C. Hooijkaas, (1998) *RESTWAQ 2, Part II. Comparison of remote sensing data, model results and in-situ data for the southern Frisian lakes,* NRSP-2 98-08b, Netherlands Remote Sensing Board (BCRS); Programme Bureau; Rijkswaterstaat Survey Department, Delft, The Netherlands, 77 + Appendices p.

Whitlock, C.H., L.R. Poole, J. Usry, W.M. Houghton, W.G. Witte, W.D. Morris and E.A. Gurganus, (1981) Comparison of reflectance with backscatter and absorption parameters for turbid waters. *Appl. Opt.,* Vol. 20, No. 3, p. 517-522.

4. Use of remotely sensed images by SPOT in hydrologic modelling

NATALIE LORENZ, MARC VAN DIJK *and* JAAP KWADIJK

WL | delft hydraulics, P.O. Box 177, 2600 MH Delft

<natalie.lorenz@wldelft.nl>

1. Introduction

Remotely sensed Earth Observation (EO) data is often used for hydrological modelling purposes. Typical information that can be provided by remote sensing is land use type, estimates of percentage paved surface and estimates of evaporation. Within the EC-funded NOAH (New Opportunities for Altimetry in Hydrology) project we focus on an alternative use, which is the use of altimetry data derived from satellites images as well as from air photographs. The altimetry data was used for risk mapping, hydraulic modelling as well as for hydrological modelling. In this paper we focus on Digital Elevation Models (DEM) in hydrologic models for sub basins of the Mosel and in the entire Mosel basin.

The study included three main tasks:

1. Stream network deliniation from Digital Elevation Models (DEM) derived from remotely sensed EO;
2. Derivation of Unit hydrographs from geomorphological characteristics with DEM;
3. Application of the unit hydrographs in conceptual hydrological rainfall runoff models to simulate discharge time series.

A general overview of the applied concept and the use of data in this study is shown in Figure 1.

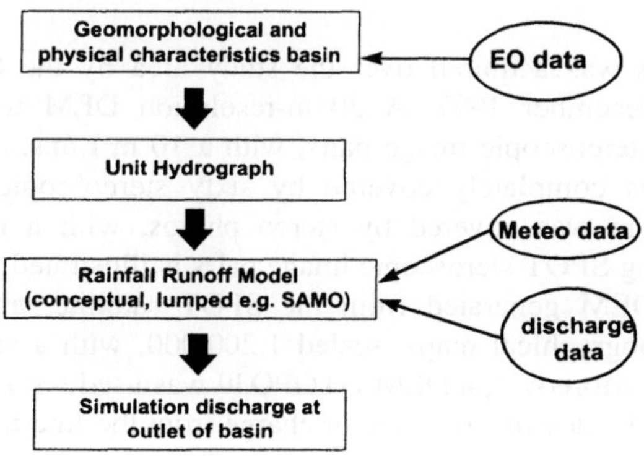

Figure 1. Overview of concept in study and use of data.

A. van Dijk and M. G. Bos (eds.), GIS and Remote Sensing Techniques in Land- and Water-management, 39–53.

This paper focuses firstly on the network deliniation from digital terrain models in the Mosel and Blies basin. Secondly the paper presents the methodology followed to apply a Unit Hydrograph derived from measured data in a larger basin (Upper-Blies basin) and in a small basin (Bliessen) using EO data.

2. The study area

The study is carried out in the Mosel basin. The Mosel is the largest tributary to the River Rhine, with a catchment area of 28,380 km^2 in France, Luxembourg and Germany. Its length is 520 km reaching from the Col de Bussang (735 m a.m.s.l.) in the Vosges mountains, France, to its outlet in Koblenz, Germany (59 m a.m.s.l.). The Vosges mountains are an ancient massif covered with forests, with a glacial relief and deep valleys. Downstream, the sedimentary plateau is an undulating agricultural landscape with some sparse forest. Annual rainfall is approximately 1000 mm, without a clear maximum during the year. Rainfall varies in the basin due to orographic effects.

The Mosel has a very variable discharge, with low flows during summer and often high floods in the winter period. The following data illustrate the river regime:

- maximum flood discharge 3,950 m^3/s (December 1993);
- mean discharge 282 m^3/s, (1931 to 1996);
- minimum discharge 27 m^3/s (July 1976).

Within the Mosel basin different sub basins were chosen to carry out the research. These were the Upper Blies basin with a sub catchment, the Saar basin and the entire Mosel basin. The Upper Blies basin (320 km^2) was selected to apply the hydrological modelling concepts and the tools utilised in this study, the Saar and Mosel basin were used to investigate the effects of scale on network delineation and application of hydrological modelling concepts.

3. Altimetry data

Remotely-sensed data was acquired over the study area by the SPOT EO satellite between May and December 1997. A 20 m-resolution DEM was generated from SPOT panchromatic stereoscopic image pairs, with a 10 m r.m.s. altimetric accuracy. The Mosel basin was completely covered by sixty stereoscopic pairs (Figure 3). The Upper Blies basin was covered by stereo photos, with a resolution of 5 m. The method of deriving SPOT steroscopic image pairs is illustrated in Figure 4.

Apart from the DEM generated from the SPOT satellite, an alternative DEM was derived from topographical maps, scaled 1:200,000, with a spatial resolution of approximately 100 m. Moreover, a DEM GTOPO30 was used with a spatial resolution of 1000 m, which can be downloaded free of charge from the Internet.

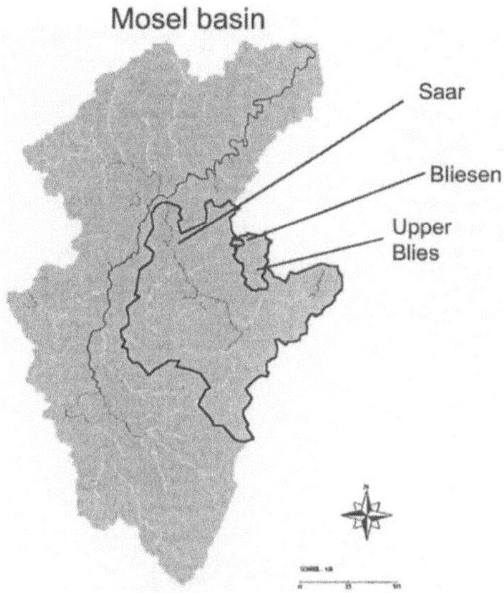

Figure 2. Mosel basin with sub catchments.

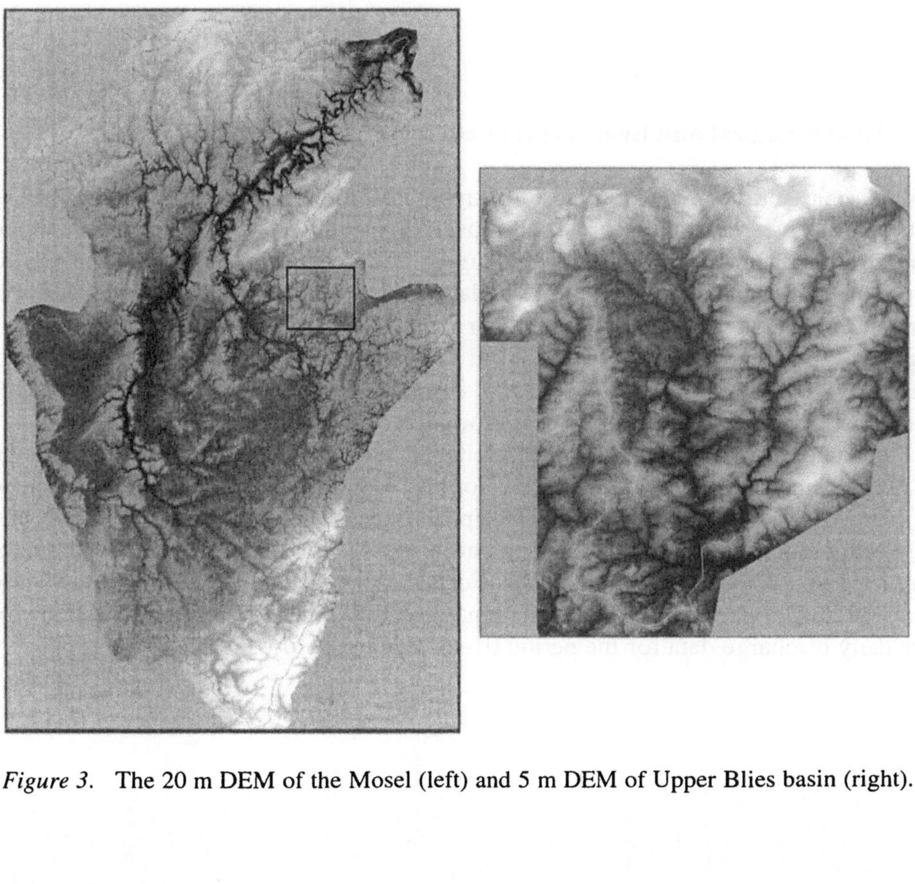

Figure 3. The 20 m DEM of the Mosel (left) and 5 m DEM of Upper Blies basin (right).

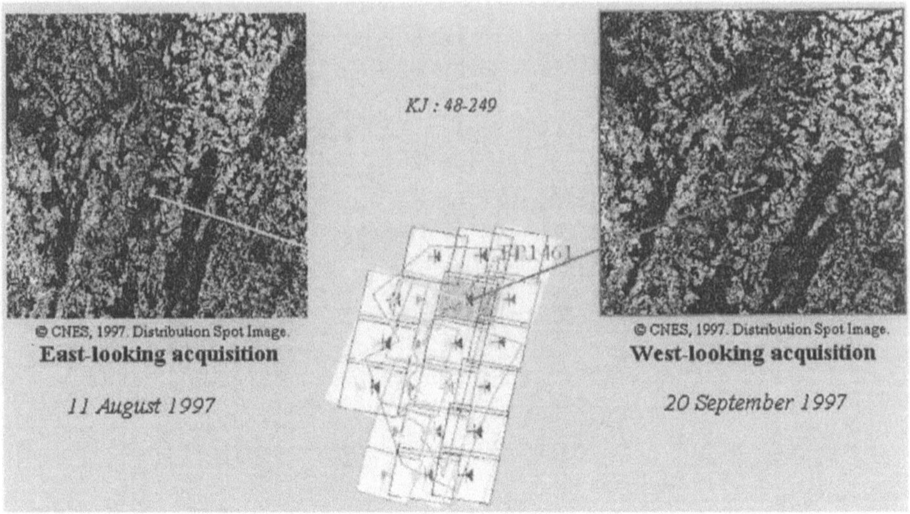

Figure 4. Derivation of steroscopic image pairs.

4. Meteorological and hydrological data

Daily discharge time series of almost 80 stream flow gauging stations on the Mosel and its tributaries were available for periods covering 8 to 30 years. Rainfall measurements were also available for more than 200 gauging stations within the French and German part of the basin, taken from an hourly to daily time resolution. Evaporation data for the German part of the Mosel was present for more than 10 stations for a period ranging from 6 to 20 years. All these hydrological and meteorological time series were stored in HYMOS (Crebas, 1994). The HYMOS system is a time series oriented information system with common facilities for data processing and (spatial) analysis.

For each study area, the areal precipitation and evaporation are calculated with HYMOS using the Thiessen method. For the Upper Blies basin, 10 precipitation stations are selected in or just near the Upper Blies basin. For the Upper Blies basin daily evaporation data from two stations are used. In the other basins several stations were used to calculate both areal series. For the Upper Blies, the Saar and the Mosel, discharge series were available for stations near the outlet of the basins. All stations have daily discharge data for the period 01-01-1990 to 01-01-1996.

5. Deliniation of drainage patterns from DEMs

Digital Elevation Models offer the possibility to automatically extract basin characteristics by creating a Local Drain Direction Map (LDD map) which gives the flow direction on each cell of the DEM. The GIS tool PCRASTER (Van Deursen, 1995) was used for processing the DEMs. Generating a LDD means that for each cell of the DEM the direction of the steepest (downhill) slope is determined, which is the direction of the local drainage. The shape and size of the basin is derived from the drainage pattern maps with the location of the basin outlet. From the drainage pattern map a flow accumulation map can also be derived with for each cell the upstream basin area. The LDD map forms the basis for defining all kind of basin characteristics such as the orders of the stream network basin area, length of streams etc. Following the scheme of Strahler (1952) a first order stream is a channel that originates at a source and the highest order stream is the outlet of the basin.

Local drain direction maps were automatically generated from the SPOT-DEM, resolution 20 m; from the 100 m resolution DEM and from the GTOPO30 DEM, resolution 1000 m.

Figure 5 shows of the Bliessen basin a so-called 'ortho image', combined with a derived LDD pattern from the 20 m DEM, together with a geographical map of the area. The LDD is derived from grid cells which have an upstream drainage area of 400 by 400 m^2. Interesting to analyse is whether the derived LDD from the DEM results in a correct location of the river streams.

From Figure 5 the following conclusions can be drawn about the correct positioning of the river course:

- areas which are covered by woods will give an inaccurate positioning of streams (area 1). This can be explained by the fact that the DEM is corrected for forest areas: here the altitude is decreased by the average tree height to attain the altitude of the local ground level. This correction of the DEM leads to inaccuracies in the determination of altitude.
- for areas which are more flat, the determination of the location of the river stream is also inaccurate because of the relatively low vertical accuracy of the DEM. In Figure 5 there is little relief around area 2, therefore the LDD does not follow the real river course.
- the river course can be located more accurate in areas where there is a lot of relief. In area 3 the contour lines on the geographical map are close to each other, which means that locally there is a lot of relief. Here, the river course is corresponding well by the drainage pattern of the LDD.

For the entire Mosel basin LDD maps were extracted from the 100 m DEM and the 1000 m DEM GTOPO30 (Figure 6). The automatically extracted drainage pattern from the 1000 m DEM (left) did not correspond to the real drainage pattern in the Mosel basin (right). Using the 100 m DEM the real pattern was well reproduced.

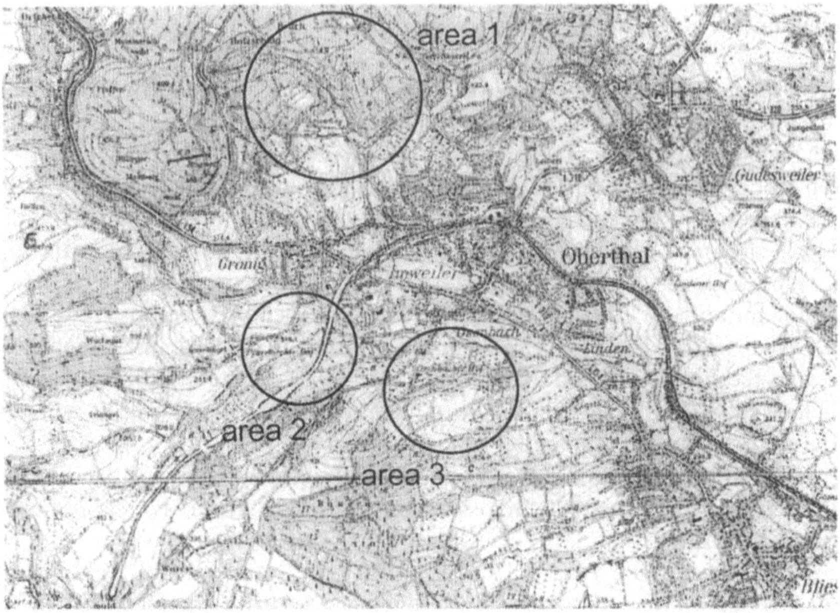

Figure 5a. Geographical map of Bliessen basin.

6. Derivation of unit hydrographs

The basic assumption of the Unit Hydrograph (UH) (Sherman, 1932) is that a basin always responds in the same manner on a certain amount of rainfall, given that the rainfall is homogeneous over the basin. This assumption implies that the response of the basin is determined from geomorphological characteristics, which do not change in (short) time. Potentially, geomorphological characteristics can be derived from DEMs.

The main principle in our research is that typical basin characteristics, as slopes, drainage pattern and basin shape and area, which are obtained from EO data, are incorporated into a Unit Hydrograph. The methodology followed is that as a first step the Unit Hydrograph is derived from measured *in situ* time series of river discharge in combination with precipitation data. As a next step we determine a Unit Hydrograph from EO data and compare this hydrograph with the hydrograph derived from measured time series.

Several methods can be applied to derive a UH from altimetry data. In this study we used the method proposed by Clark (1945). The method of Clark, also known as the *Time-area histogram method*, aims at developing an Instantaneous Unit Hydrograph (IUH) due to an instantaneous rainfall excess over a basin. It is assumed that the rainfall excess first undergoes pure translation followed by attenuation. The translation is achieved by a travel time versus area histogram and the attenuation by routing the results of above through a linear reservoir at the basin outlet. The description of the Clark's method is described well in Subramanya (1994).

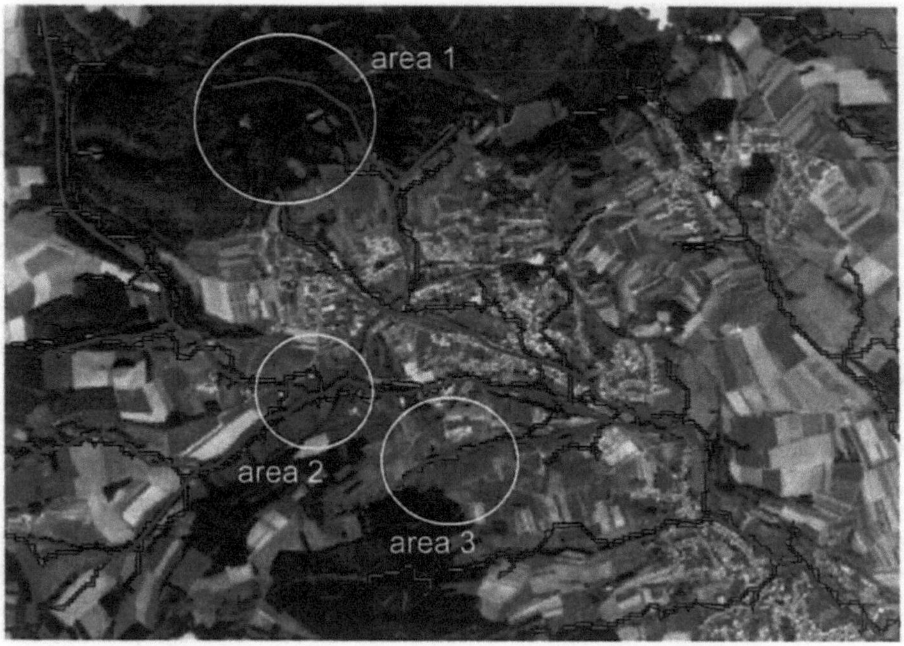

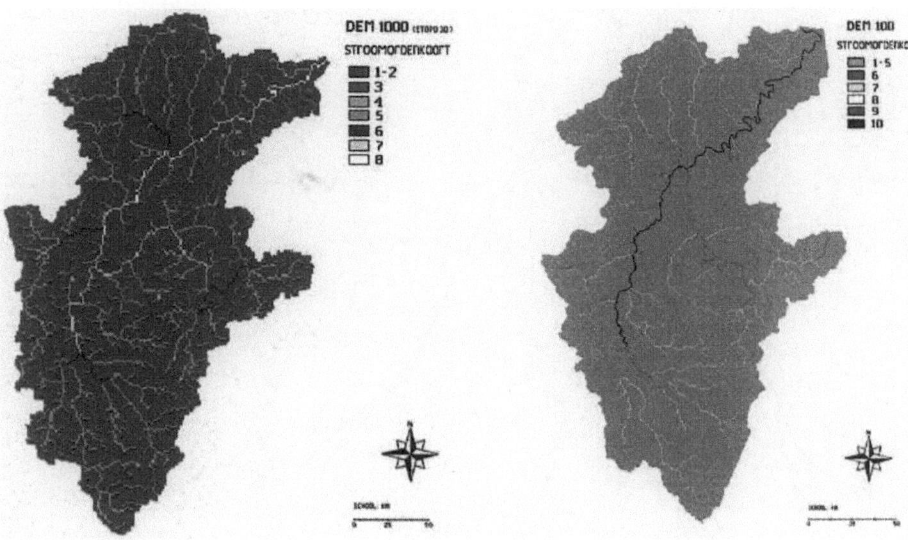

Figure 6. LDD maps of Mosel basin from GTOPO30 1000 m DEM (left) and the SPOT 100 m DEM

Time refers here to the time of concentration, which is the time required for a unit volume of water to travel from the farthest point of the basin to the outlet. Consequently the total basin area drains into the outlet in T_c hours. The basin can be divided in N sub-areas that have an equal time interval of $T_c/N = \Delta T_c$. The inter-isochrone areas A_1, A_2, ...A_N are used to construct a travel time-area histogram. In Clark's method the river basin is assumed to be a linear reservoir, described by $S = K*Q$ at the outlet, with K as the *storage time constant* (unit: time). This constant is estimated from recession limbs of measured runoff hydrographs. Knowing K of the linear reservoir, the inflows at various times are routed by the Muskingum method (with $x = 0$, because of linear reservoir) (McCarty, 1938). The routing of the time-area histogram results in the ordinates of the Instantaneous Unit Hydrograph (IUH) for the basin. Knowing the IUH, the 1-hour UH can be derived.

6.1. UH from measured time series

For the Blies basin the average rainfall series is screened for storms. Individual storms, useful for the derivation of a UH, are of fairly uniform rainfall intensity and a rather restricted duration. Furthermore the rainfall events should be selected from winter periods, because in summer the relative rainfall loss is too high and the derived UH can therefore be inaccurate. Eventually three storms with hydrographs are selected from the series, indicated by the date of the peak discharge: 01-03-1990, 11-01-1995 and 19-03-1995 (Figure 7).

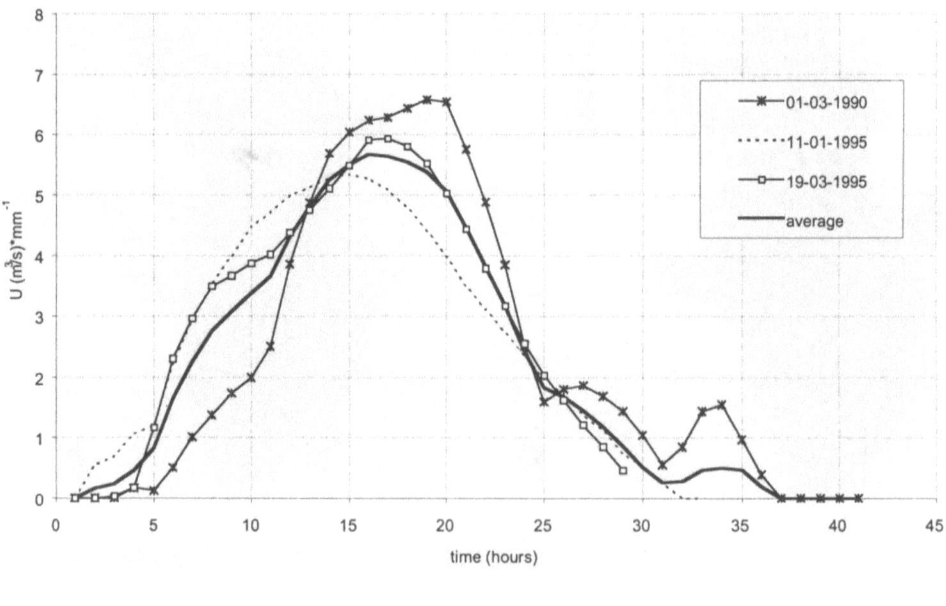

Figure 7. Derived 1-hour unit hydrographs from measurements.

The recession constant can be derived from the recession limbs of measured hydrographs. The recession limb can be divided in three stages of storage depletion: surface storage, interflow storage and groundwater storage (i.e. base flow storage). The recession constant which is necessary for the method of Clark is the recession constant for surface storage, K_{rs}. The recession of a storage can be expressed as:

$$Q_t = Q_o K^t_{rs} \qquad [L^3/T]$$

in which Q_0 and Q_t are discharges at a time interval of t days with Q_0 being the initial discharge. K_{rs} is expressed in days, and is less than 1. The typical value of the surface recession constant is between 0 and 0.20, and of interflow storage between 0.50 and 0.85. The measured hydrograph is plotted on semilog paper, and a straight line is drawn along the beginning of the recession limb.

6.2. *UH from EO data*

The procedure to derive an IUH from GIS data, combined with the method of Clark, is as follows:

- The DEM with a resolution of 20 m is entered into PCRASTER (Van Deursen, 1995), from which a LDD map (local drain direction) is derived. After all the pits (depressions) are removed in this map (filling up of the cores), a single pit is left over: the outlet of the river basin.
- From this LDD map, a distance map is made in which the distance from every cell to the point of outflow is known.
- This distance map is combined with a *velocity map*, in which the velocity of the water is given for every cell. The procedure to derive this map is described below.
- By combining these two maps, a travel time map is created. For every cell the travel time from this cell to the outlet is calculated. The maximum travel time calculated is the *time of concentration Tc*.
- From this map a time-area histogram is made.
- The storage time constant/recession constant K is derived from measured hydrographs.
- With K the time-area histogram is routed, calculated in a spreadsheet, which results in the ordinates of the Instantaneous Unit Hydrograph (IUH).
- From the IUH the 1-hour UH is calculated, by lagging the IUH by 1 hour and dividing the ordinates by 2.

6.2. *Velocity maps*

Initially the velocity is assumed to be *constant* over the entire catchment. Comparing the average measured unit hydrograph with the unit hydrograph derived with a constant velocity field, the velocity of 0.7 m/s leads to the best results. Methods that use distributed velocity fields (Maidment et al., 1996) and a method developed during the research did not provide better results (Figure 8).

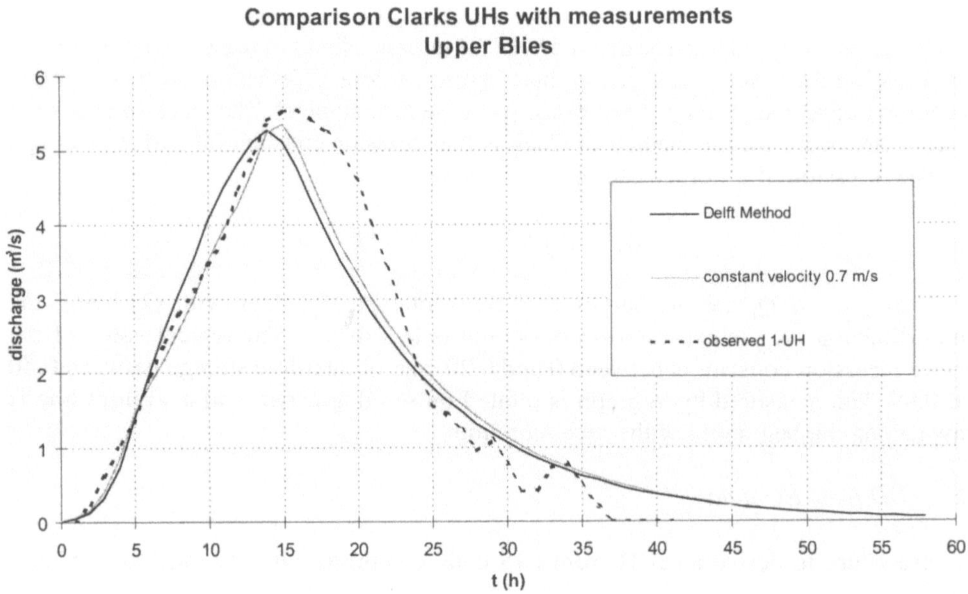

Figure 8. Comparison of Clarks UH with the measurement, with constant and distributed velocity field

7. The sacramento model

In this study we used the SACRAMENTO (SAMO) model (Burnash et al., 1973) as a conceptual rainfall runoff model. In the simulation of the runoff process by the Sacramento model a distinction is made between the *land-phase* and the *channel-phase*. The *land-phase* component of the Sacramento model is divided into the pervious and impervious part of the basin. From the *impervious* areas, precipitation immediately discharges to the channel. However, impervious areas, which drain to a pervious part before the water reaches the channel, are not considered as impervious. The drainage system of the *pervious* part (generally the main part) of the basin is divided into:

- an *upper zone*, representing the basin surface system;
- a *lower zone*, representing the basin groundwater reservoir system.

Both zones have a tension and a free water storage element. Tension water is considered as the water closely bound to soil particles. Generally, first the tension water requirements are fulfilled before water is entering the free water storage, although some important exceptions are present.

8. Application of UH from EO data in hydrological modelling

8.1. *Sacramento model of Upper Blies Basin*

The unit hydrograph derived from Clark, with the constant velocity field, is applied as unit hydrograph in the SAMO-model. The calibration has been carried out by a sensitivity analysis of the parameters versus the calculated discharge. During this process we concentrated on the winter periods and the peak discharges, regarding the objective of simulating floods. The model results for the validation period are given in Figure 9. For the extremely high peak discharges, the simulated discharges are little too low, for dry periods the simulated discharges are a little too high. The overall results are satisfying.

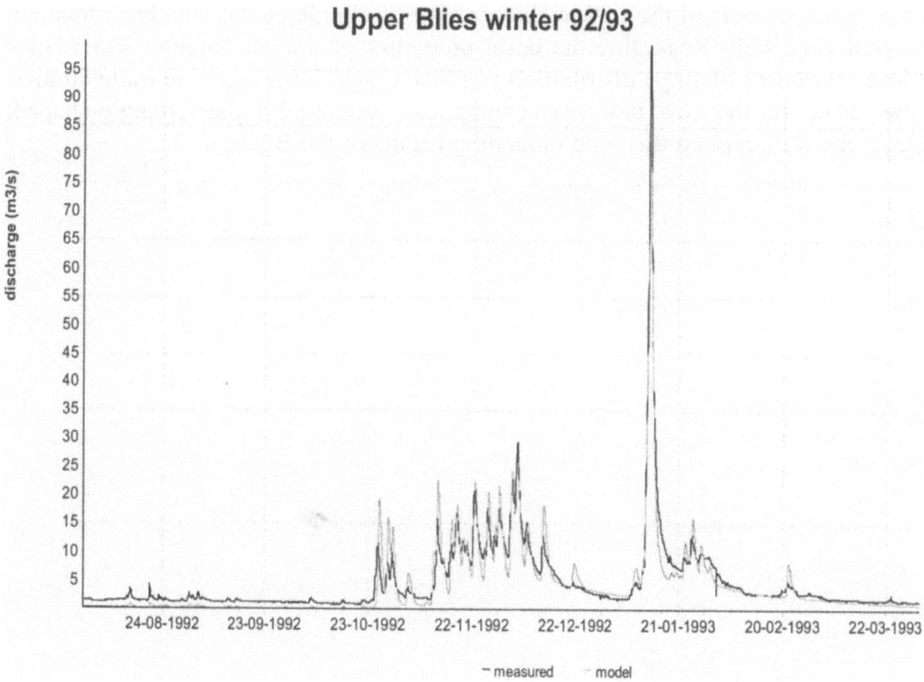

Figure 9. Discharge simulation in validation period of Upper Blies with Sacramento.

8.2. *Application Sacramento model for the Bliesen sub basin*

By using the DEM of a larger basin, it is also possible to delineate any sub basin within this basin. Often larger basins are gauged at their outlet, but not within the basin area. Therefore it is interesting to analyse the possibilities of using a calibrated model of a larger basin in combination with the DEM to obtain a discharge model of a smaller sub basin. This is however only applicable to comparable catchments, in terms of distribution of slope and of land use. In this section we will analyse the possibilities of using the Sacramento model, calibrated on the Upper Blies (321 km^2), for modelling the discharge of a much smaller sub basin called 'Bliesen' (17.1 km^2).

For the Bliesen we derived a UH using Clark similar to the one of the Upper Blies. To be able to obtain the UH with Clark, an estimate of the recession constant K is necessary. From the DEM the time of concentration can be derived for every sub-basin, using the same field velocity. In this time of concentration lies the 'reaction time' of the catchment, influenced by the drainage pattern, the distribution of slopes and land cover, all basin properties which determine the surface runoff. Therefore the recession constant for surface storage K will also be related to the time of concentration. Consequently, the recession constant of surface storage for the Bliesen is determined linearly from the two time of concentrations of both basins.

In Figure 10 the results of the obtained Bliesen model are compared with the discharge measurements of the station Bliesen. The discharge of this smaller catchment is simulated very well. Regarding the basin properties of the catchments Upper Blies and Bliesen (in terms of slope distribution and land cover) both areas are quite similar. Also the shape of the two basins is comparable: both are rather triangle-shaped. This may very well explain the good modelling results of the Bliesen.

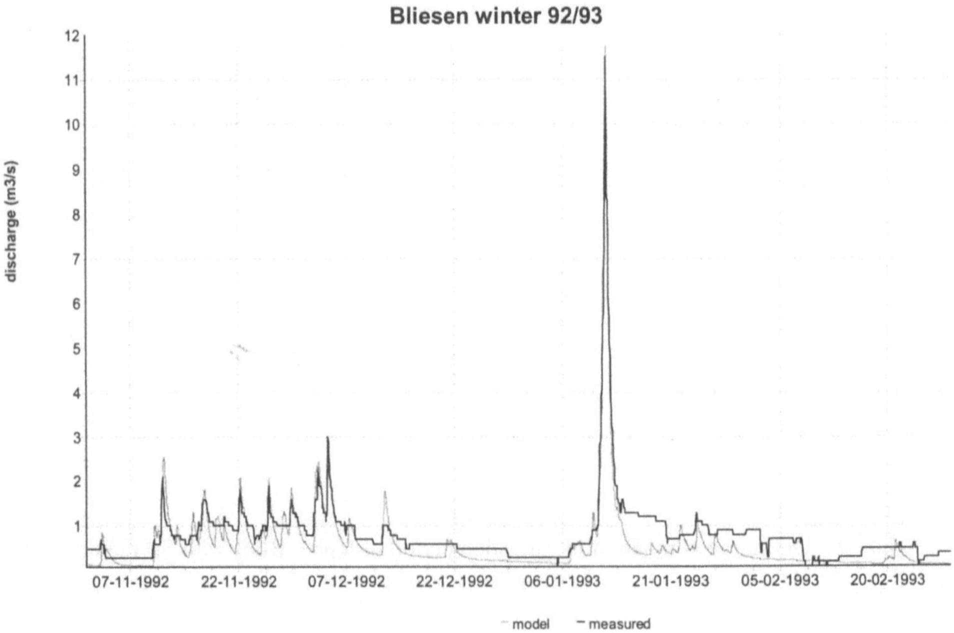

Figure 10. Discharge model obtained from downscaling calibrated Upper Blies model to smaller catchment Bliesen

9. Costs and benefits from spot data

We evaluated the time needed to use EO data and GIS for hydrological modelling versus the time consumption of the method done by hand, specified for the derivation of the Clark Unit Hydrograph. This evaluation is possible using experience of WL | delft hydraulics in a project for the Indus in Pakistan (National Engineering Services Pakistan, 1997). The total process of deriving a unit hydrograph with the method of Clark is subdivided into several activities together with the required time and the results are presented in Table 1. From this comparison it was concluded that using the method of Clark combined with GIS instead of ordinary geographical maps results in gaining of time of circa 80%.

Table 1. Comparison time consumption of method of Clark by hand and with the use of EO data and GIS

Method Clark by hand	
Activities	time (day)
Collection topographical maps with river network of main streams	*
Collection of precipitation data and measured discharges/water levels/water level-discharge relations	*
Collection of field data: cross and length profiles, indication of frictions	*
Subdividing main streams in 'equivalent' stream sections	0.25
Determination of travel times in main streams of equivalent sections	2
Drawing iso-lines in entire catchment on basis of estimated velocities from field data	3
Carrying out planimetry	0.75
Derivation recession constant K from measurements	*
Derivation A-t-diagram by speadsheet	*
Total	6.0

Method Clark with EO data and GIS	
Activities	time (day)
Collection DEM of sufficient accuracy	*
Collection of precipitation data and measured discharges/water levels/water level-discharge relations	*
Collection of field data: cross and length profiles, indication of frictions	*
Adapting resolution in order to achieve manageable computer processing time	0.25
Derivation of basin area and LDD with GIS	0.25
Checking the basin area and the LDD on failures	0.25
Derivation of distance map, applying velocities, estimated from field data in main streams, applying on rest of catchment constant flow velocity	0.25
Derivation recession constant K from measurements	*
Derivation A-t-diagram by speadsheet	*
Total	1.0

10. Conclusions

The concept of using SPOT EO data for the derivation of a UH, in combination with a rainfall-runoff model, enables quick and reliable simulation of discharge time series. Of course, the reliability depends also on the quality of the input data, such as rainfall and evaporation data, sufficient spatial distribution of measurement stations, and the quality of the DEM.

When using the method of Clark, at least one measured hydrograph at the basin outlet, with a considerable contribution of direct surface runoff, has to be available to know the storage constant of direct surface runoff. If a hydrograph is not present, information about the *possible* value of the storage constant of surface runoff may be derived from the larger basin, of which the smaller studied basin is a part, by the use of EO data. The storage constant can be downscaled linearly using the time of concentration of both basins, derived from EO data and GIS. This method however can only be applied to two basins of the same shape and distribution of slope and vegetation/land use.

In this study we evaluated the time consumption of using EO data and GIS for hydrological modelling versus the time consumption of the method by hand. This comparison is specified for the derivation of the Clark Unit Hydrograph. In this evaluation we used experience of WL | delft hydraulics in a project for the Indus in Pakistan (National Engineering Services Pakistan, 1997), where Clark was applied and carried out by hand. The total process of deriving a unit hydrograph with the method of Clark is subdivided into several activities together with the required time. This comparison shows that 80% of the time can be saved by using the method of Clark combined with GIS instead of using ordinary geographical maps, and carrying out the method by hand.

The results of the hydrological modelling in the NOAH project show that SPOT EO data provides a useful addition on the current use of EO data in hydrology as this data enables to produce very detailed Digital Elevation models. Using these elevation models basin delineation can be carried out and geomorphological basin characteristics can be determined.

The resolution of the SPOT EO data is particularly useful for hydrological modelling in small basins. For larger basins the amount of information is so large that the data handling with the current generation of Personal Computers becomes very time consuming. Also, the cost to cover large basins with SPOT stereo pairs is high, because the required resolution for good hydrological modelling in large basins is less than is provided by the SPOT EO data.

Application of EO data in hydrological modelling is particularly useful in areas with few available data sources and with low resolution geographical information. In well measured areas such as the Mosel basin, the interest of EO data is particularly in small sub basins, for which relatively little information is available. As the vertical resolution of the SPOT data is less accurate, this data is less useful for the application in hydrological models in flat basins.

11. Acknowledgements

We would like to express sincere thanks to the International Commission on the hydrology of the River Rhine (KHR/CHR), the Landesamt für Umweltschutz, Saarbrücken and the Deutscher Wetterdienst, Offenbach for their support in advice and data.

The European Commission is greatly acknowledged for funding the project.

12. References

Burnash, R.J.C., *et al.* (1973) A generalized streamflow simulation system. Conceptual modelling for digital computers. Dept. of Water Resources, Sacramento.

Clark, C.O., Storage and the Unit Hydrograph (1945) Trans. Am. Soc. Of Civ. Engineers, Vol. 110, pp. 1419-1446, paper No. 2261.

Crebas, J.I. (1994) HYMOS: a database management and processing system for hydro-meteorological data. Paper presented at the First International Conference on Hydro-informatics, Delft, the Netherlands, September 19-23.

Maidment, D.R. *et al.* (1996) Unit Hydrograph derived from a spatially distributed velocity field, Hydrological Processes, Vol. 10, 831-844

McCarthy, G.T. (1938) The Unit Hydrograph and Flood Routing, Conf. North Atlantic Div., U.S. Corps of Engineers, New London, Conn.

National Engineering Services Pakistan and Delft Hydraulics (1997) Development of Indus Flood Forecasting system, Final Report

Sherman, L.K. (1932) Streamflow from Rainfall by the Unit-graph Method, Eng. News Record, vol. 108, pp. 501-505

Strahler, A.N. (1952) Dynamic Basis of Geomorphology, Geol. Soc. Am. Bull., vol. 63, pp. 923-938.

Subramanya, K. (1994) Engineering Hydrology. Second edition.

Van Deursen, W.P.A. (1995) Geographical Information Systems and Dynamic Models; development and application of a prototype spatial modelling language, Thesis, NGS 190, Knag/Faculteit Ruimtelijke Wetenschappen Universiteit Utrecht.

5. Measuring and analysing of flood waves in Rhine river; an application of laser altimetry

ARDIS E. BOLLWEG *and* RENÉ VAN HEERD
Survey Department of Rijkswaterstaat, Section Remote Sensing and Photogrammetry
Postbus 5023; 2600 GA Delft
<a.e.bollweg@mdi.rws.minvenw.nl> and <r.m.vheerd@mdi.rws.minvenw.nl>

Abstract

The Survey Department of Rijkswaterstaat in The Netherlands makes intensive use of laser altimetry for topographic measurements. A new application is to measure the water level, during flood waves in rivers. As the laser signal reflects differently on water compared to land, a high point density was used. This large data set gives spatial detailed information about the flood wave, the along — and across — track slope of the river and information to derive the water resistance on different vegetation areas.

The laser data gives a high detail of information, and the breakwaters in the river are clearly visible. Computing the different roughness coefficient with laser and additional discharge measurements, will be an important improvement for water management. These roughness values will optimise the hydraulic models used to predict the water movement in rivers.

In this paper the laser altimetry technique is being explained, and some examples are shown. Then the Rhine project will be discussed, and a first attempt to give an indication of the roughness coefficient are shown.

1. Introduction

The Survey Department of Rijkswaterstaat in The Netherlands has been investigating the feasibility of obtaining topographic information from laser measurements since 1988. Starting with the laser profiler, this technique developed rapidly to a well used technique to measure heights and elevations. In 1996 the Survey Department started with a project (Recent Elevation Model of the Netherlands, called in Dutch "Actueel Hoogtebestand Nederland" (AHN)), to measure the elevation of the whole country with laser altimetry every 4 x 4 meters. The present situation of the AHN-project is shown in Figure 1.

As the years went by, we obtained more knowledge on the laser technique (Huising, 1998) and error analysis (Lemmens,1997). But also new ideas on applications were researched (Wouters, 1998). A new application is to measure the water level in rivers. In this paper we will give an impression of the application to measure the elevation of flood waves in rivers by laser altimetry. This high spatial data gives

A. van Dijk and M. G. Bos (eds.), GIS and Remote Sensing Techniques in Land- and Water-management, 55–67.

detailed information about the slope of the water along track and cross track in the river, and the flow velocities, for the input of hydraulic models of rivers.

First the laser principle will be explained, examples of the Recent Elevation Model project (AHN) will be shown, and an example of water management by laser will be discussed.

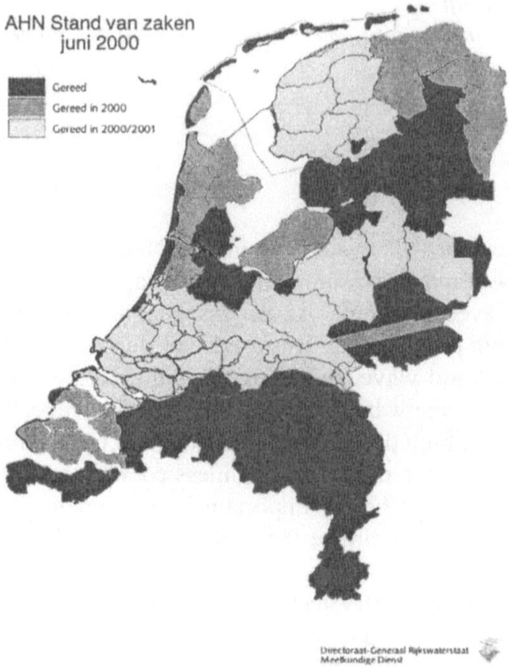

2. Laser altimetry principle

An airborne laser scanner system uses three main instruments, a laser instrument, a Global Positioning System receiver and an Inertial Navigation System (to measure pitch, roll and heading of the platform). This laser-system is mounted on a platform and installed in an aeroplane or helicopter. By using a ground reference GPS-station, and the data acquisition software, the X, Y and Z co-ordinates can be determined, see Figure 2.

The basic principle of the laser system is to compute the distance between laser and a point by measuring the time between transmission of a laser pulse and the return signal, after a hit on the ground. By computing the position and orientation of the platform by the GPS and INS, the position of the footprint (the point where the laserbeam hits the earth) is known, with a height precision of approximately 20 cm standard deviation.

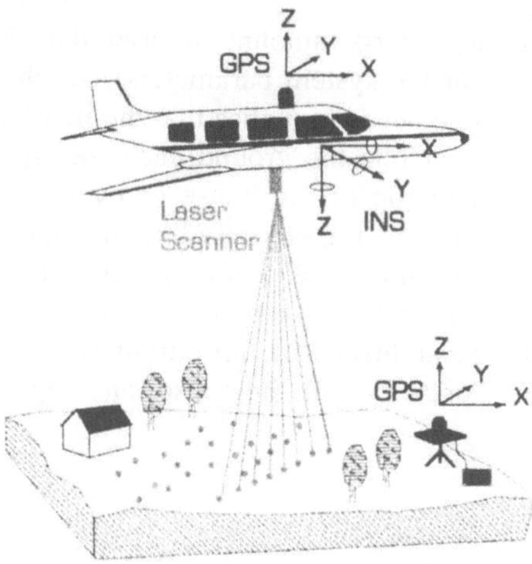

Figure 2. Principle of laseraltimetry.

The Survey Department uses laser altimetry data for different projects. The main project is AHN, executed in co-operation with the water boards and provinces of the Netherlands. But also for coastal management, road planning and design, and river management, laser altimetry data is used. For all projects different laser-systems are used. In this study, an application for water management, the Saab laser scanning system is used, see Table 1 for the specific instrument parameters. This system, used in an aeroplane, is operating at an average height of 480 m. This height is convenient, because often the clouds start at a height of 500 m, see Table 2. A laser system can operate during day and night, but not when there is rain, smog, low clouds, snow or strong winds.

Table 1. Parameters of the Saab laser scanning system.

Parameters	Value
laser system	TopEye
flight altitude	480 m (average)
aircraft speed	55 m/s (average)
wavelength laser	1064 nm
scan angle	+/-20^0
scan/mirror frequency	16.67 Hz
pulse rate	5260 Hz
scan spot size/diameter footprint	0.48 m
divergence signal	1 mrad
swath width	350 m
spots per swath	315
line spacing	100 m
strip overlap	60% overlap = 210 m
average point density (in one flight)	1 point /3.7 m^2
average point density (for whole area)	1 point / 1 m^2

With this laser technique a large amount of laser data is obtained, with a high point density (see Table 1 for the system parameters). As the laser just measures the height within the footprint, one needs to realise that the measured point is not a typical topographic point. As the laser hits the ground each meter, the return signal can be ground level or water level or top of the house or even a tree or a bird. The laser data, obtained from the platform are processed to the national co-ordinate system (for the Netherlands the 'RD' and 'NAP' system) and then reduced for blunders, i.e. measurement errors or measurements on clouds, or even birds. This data set can then be filtered for unwanted measurements, i.e. on buildings and vegetation. In this study the data used is unfiltered. For AHN-data the houses and vegetation are filtered.

3. AHN applications

The use of the AHN data is not only for the demand for detailed information about elevation, from water boards, provinces and the national government. These institutions require this information for the management of coasts, dykes, polders and higher-level areas in The Netherlands which seem to be drying out, for an example see Figure 3. But also for the construction of infra-structural works the data are of great importance. The elevation information is used in models that simulate the discharge of the rivers, or that simulate the effect of a dyke breach. But also flight simulators or even virtual reality will use this data. Other application are the noise pollution caused by the road affected on high rise flats, and computing the optimal position for a transmitting-station for telephone companies. Last but not least the application for water management, discussed in the next section.

Figure 3. An unfiltered elevation data set of 5 by 3 kilometres with a point density of one point per 16 square metres. The image shows a shaded relief of a Dutch village Klundert in Noord-Brabant. The houses, forest, dykes and water areas are clearly recognised.

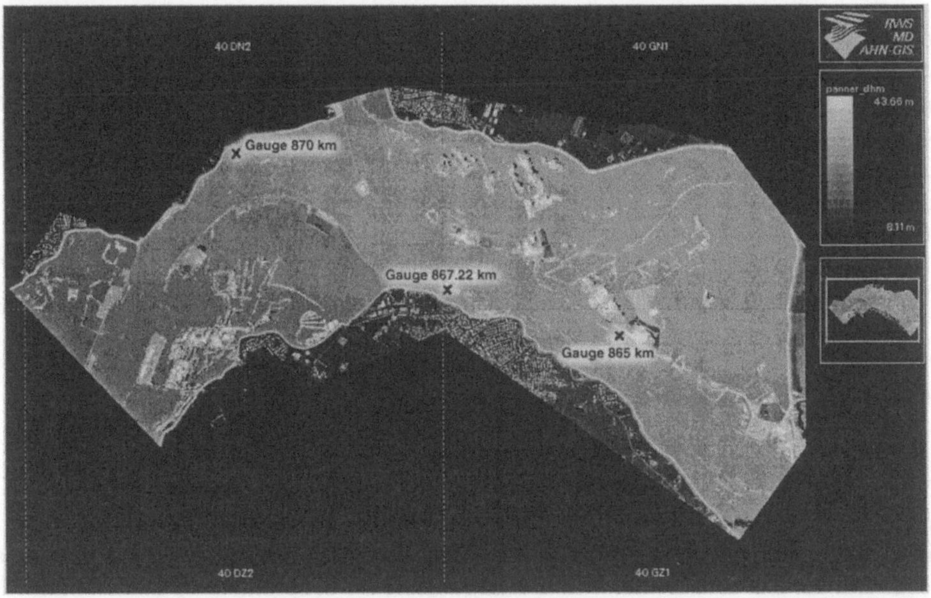

Figure 4. The high water area (9 by 4 km) of the Rhine at 5 and 6 November 1998, with an average point density of 1 point per square meter. White is high level, black is low, or no data.

4. Project water level of the Rhine river

At the beginning of November 1998 a flood wave was approaching the Netherlands via the Rhine. To be able to measure water levels, the volume of the high water level situation and the slope of the river, a laser altimetry acquisition was carried out on the 5th and 6th of November by the aerial survey company Eursosense, using the Saab laser scanning system, see Figure 4. This test area is approximately 9 by 4 kilometres and covers the flood plain of the Rhine near "Pannerdensche Kop". This area was chosen because of the specific variable conditions of the bottom topography of the river.

5. Point density

As a laserpuls reflects differently on water compared to land due to the smooth surface of water, the acquired point density was related to the water-return signal and the swath width. For this application a homogeneous point density was acquired above water, which resulted in a point density of 1 point per square meter.

The point density depends on the speed and height of the aeroplane, and the swath width. Indirect these parameters are affected by the weather conditions, mainly wind-speed and -direction. For the two measurement days, the weather conditions are mentioned in Table 2.

Table 2. Weather characteristics of KNMI in Deelen.

weather characteristics Deelen*	5 Nov. 16:00 1998	6 Nov. 9:00 1998
wind speed (m/s)	8	6
wind direction (deg.)	230 (ZW)	270 (W)
cloud coverage	7/8	6/8
height of first cloud layer (m)	540 (cumulonimbus)	600 (cumulonimbus)
height of second cloud layer (m)	750 (cumulonimbus)	1680 (stratos nimbus)

* The weather station Deelen is located 23 kilometres to the North-West of the centre of the project area.

The point density in a single flight above water or land is 1 point per 3 to 5 m^2, dependable on speed of the platform and drift. For the transition from land to water, straight underneath the platform, only a slight decrease (1%) in received signal is noticed. But for the total swath width a large decrease in returned signal is obtained, 25 to 40%, see Figure 6. This can be explained by two effects. First the reflection of the water surface, which reflects the signal in the same angle as received, especially for low angle signals. The swath width decreases with approximately 30 to 50% compared to the swath width on land, see Table 1. Second effect is that at the edge of the swath strip, where the point density is higher, due to the pendulum movement of the scanner, see Figure 5a and b. Both effects could cause the decrease in returned signal. Another reason for decreasing of the swath width is due to the drift. In the next section this will be explained, by reconstructing the flight parameters.

6. Flight parameters

Using the laser data of one flight the measurement situation can be reconstructed. The time needed to fly from the east to west on flight number 27 (this flight is twice measured, on 5 and 6 November 1998) takes 212 seconds on November 5th. The true length straight under the aeroplane is approximately 9240 meters, which gives a ground velocity of 43 m/s. The direction of the flight on the ground is 280 degrees to the east computed from the north direction. The scan angle on this flight using an oscillating scanner is 348 degrees to the east from the north direction. From Table 2 we see that the wind is from 230 degrees. The real flight angle, meaning the direction of the platform, is perpendicular to the scan-lines, and can be computed at 258 degrees. Due to the wind a drift is seen in the laser data of 22 degrees, see Figure 7.

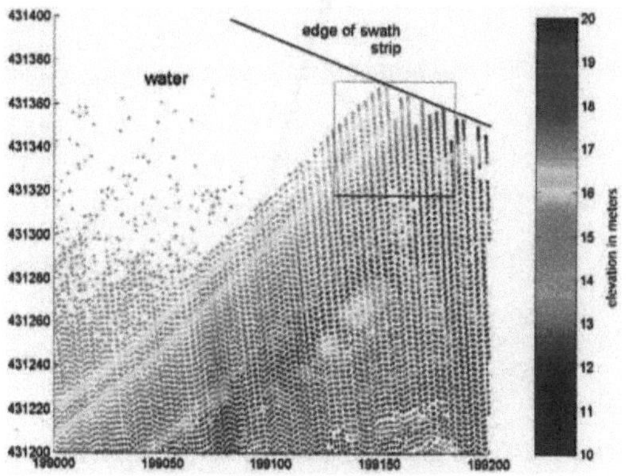

Figure 5a. Difference in point density on land and water. Left of dyke is water, on the right side is land. Co-ordinates in meters.

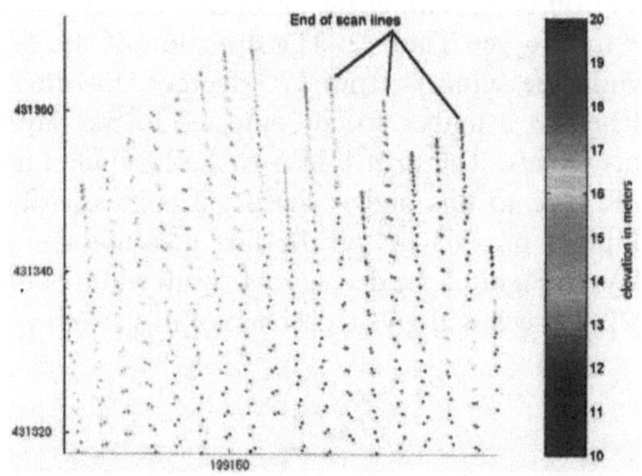

Figure 5b. Point density on land at edge of swath strip. Enlargement of square in Figure 5a. Co-ordinates in meters.

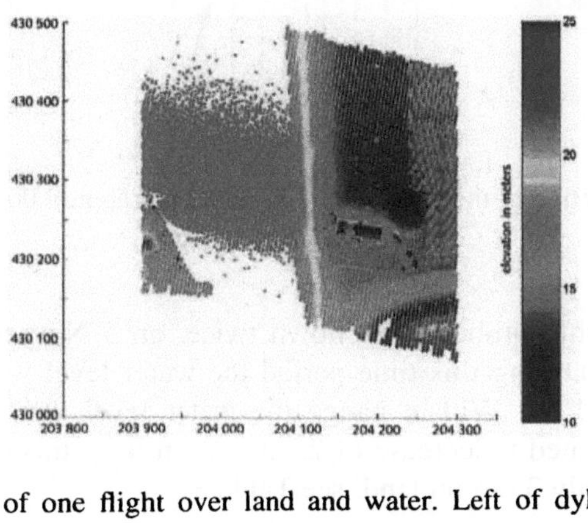

Figure 6. Laser data of one flight over land and water. Left of dyke is water, on the right side is land. Co-ordinates in meters.

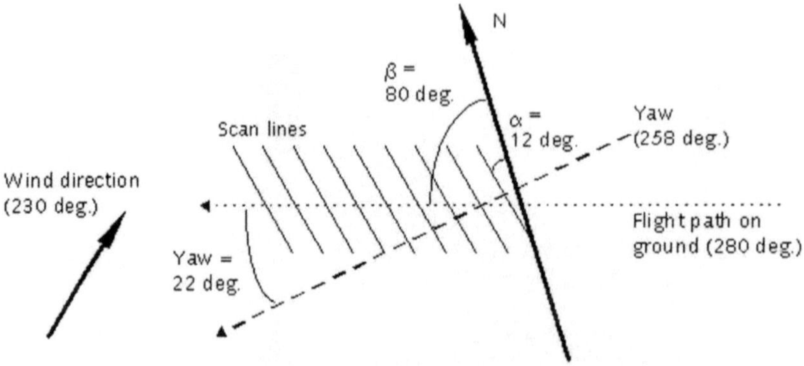

Figure 7. The reconstruction of the flight path # 27 on 5 Nov. appr.16:00 with a yaw of 22 degrees, due to the sideward wind.

For flight number 27 on 6 November the aeroplane has almost headwind, according to the KNMI weather report, see Table 2. The direction of the flight on the ground is again 280 degrees, while the wind is from 270 degrees. In order to make up for this headwind, the aeroplane has a higher speed compared to the day before, which results in a ground velocity of 77 m/s. The drift is due to the headwind almost negligible, only 4 degrees, see Figure 8. Due to this higher speed the point density is less then the day before, for example 1 point per 3.3 m^2 on the first measurement day, and 1 point per 5.7 m^2 on the second day. In Figure 7 the decrease of swath width of the flight path is shown, due to drift. A yaw of 22 degrees will give a decrease of this swath width of 7%.

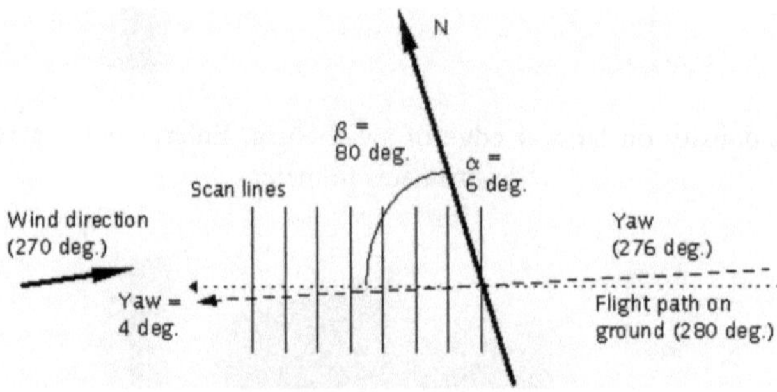

Figure 8. The reconstruction of the flight path # 27 on 6 Nov. appr. 9:00 with a yaw of 4 degrees, mainly due by headwind.

As mentioned, flight number 27 is flown twice, on 5 Nov at 16:00 hours and on 6 Nov at 9:00 hours. During this time period the water level was decreased, because the flood wave passed by. With a gauge the water level was also recorded on the same times, and computed a decrease of 25 cm. With laser this decrease in waterlevel averaged 23 cm and only 3 cm on land, see Table 3.

7. Gauges versus laser altimetry

Using the gauges along the Rhine river, it is possible to compare two independent measurements; laser and gauges. Figure 4 shows the location of three gauges along the river, from which measurements were taken (automatic and manual) of the water level.

The gauges record water levels every hour. The laser data is compared to these by using a test area of 25 x 25 m. As the laser data is distributed homogeneous in this test area, an average height point is computed over the area. This area is expected to be horizontal flat, with no slope, but some noise due to the rippling water surface. In Table 4 the laser- and gauge water levels are shown. The index *n* indicates the number of laser points in one flight in this test area. The standard deviation of the mean laser height in the test area is also indicated in Table 4. This number probably indicates the amount of rippled water.

From the differences in Table 4 one can conclude that the laser has a high precision for the averaged point. As laser gives detailed information of the area of one moment, this will be useful for computing roughness coefficients of different areas with different topography. The differences between laser and the gauges is in the order of the standard deviation of the laser in the test area.

8. Along track slope in river

During the measurement days the flood wave was passing by. As the top of the flood wave was already passed this test area 1 or 2 days ahead, the water level will decrease at all gauge stations, and in the time. To compute the along track slope, use is made of 6 gauges, of which 4 are manually and 2 automatically. In Figure 9 the decrease of slope is shown, computed by the 2 automatically gauges, approximately 5 km apart from each other. The slope variates from 14.1 cm/km to 11.3 cm/km within 2 days.

9. Waves

Instead of averaging all data within one surface it is also possible to regard the differences in the data for information. As can be noted the differences are spatially distributed. On average the variation in the water levels is approximately 30 to 40 centimetre which can be well explained as being the effect of waves within the bed. For longer waves particular waves could even be visualised on a normal white to black scale, see Figure 10. This information could give more information on the roughness coefficients used in hydraulic models.

Dependable on the roughness of the bottom topography, the water has a different discharge. With rough topography the water will have a low flow, and a higher water level is expected. By using the laser data and the computed water heights from the hydraulic model an indication on the precision of the laser data and model can be made. This can also be done with the discharge measurements and the 2D hydraulic model, which will give an idea of the water velocities in the river. Special interest will go to the roughness coefficients of specific topography to optimise the hydraulic models.

Table 3. Laser height difference between 2 flights on water and land.

Test area (100 x 100 m)	characteristics	height at 5 Nov. 16:00 (cm)	number of points[a] (n)	standard deviation[b] (cm)	height at 6 Nov. 9:00 (cm)	number of points[a] (n)	standard deviation[b] (cm)	height difference (cm)
flight # 27.4	water	1412	3139	6	1393	1705	8	19
flight # 27.8	water	1525	2892	6	1499	1549	7	26
flight # 27.9	water	1512	3301	4	1490	1738	3	22
flight # 27.10a	water	1514	3056	4	1488	1735	3	26
flight # 27.10b	land	1306	3145	[c]54	1309	1712	[c]56	3

[a] the number of points used in an area of approximately 100 x 100 m, to compute the average height.

[b] the standard deviation of the average height from the number of points used.

[c] in this area the standard deviation is influenced by roughness of the land for example vegetation.

Table 4. Comparison of gauge and laser water heights.

Measurement time (hh:mm)	gauge 870 km (manual) water height (cm)	laser water level(2) (cm)	stand. deviation test area laser (25 x 25 m) (cm)	points in test area (25 x 25 m) (n)
8:00	1470	-	-	-
10:43	-	1461	10	26
10:52	-	1458	5	173
10:57	-	1456	5	126
11:06	-	1458	9	41
12:00	1465	-	-	-
±11:00(3)	1466	1458	1(1)	366

Measurement time (hh:mm)	gauge 867,22 km (automatically) water height (cm)	laser water level(2) (cm)	stand. deviation test area laser (25x25 m) (cm)	points in test area (25x25 m) (n)
11:57	-	1488	8	12
12:00	1489	-	-	-
12:04	-	1492	4	178
12:12	-	1492	4	127
13:00	1488	-	-	-
±12:00(3)	1489	1492	1(1)	317

Measurement time (hh:mm)	gauge 865 km (manual) water height (cm)	laser water level(2) (cm)	stand. deviation test area laser (25x25 m) (cm)	points in test area (25x25 m) (n)
12:00	1520	-	-	-
15:47	-	1516	44	43
15:54	-	1521	4	195
16:00	1515	-	-	-
16:03	-	1514	4	128
±16:00(3)	1515	1518	1(1)	366

(1) The expected standard deviation for the mean height is computed by the variance of the laser points (20^2 cm^2) divided by the number of laser points.
(2) For the water level height of the laser, an average height is used, in a test area of 25 x 25 m, close to the gauge.
(3) The gauge measurements are interpolated and the laser measurements averaged.

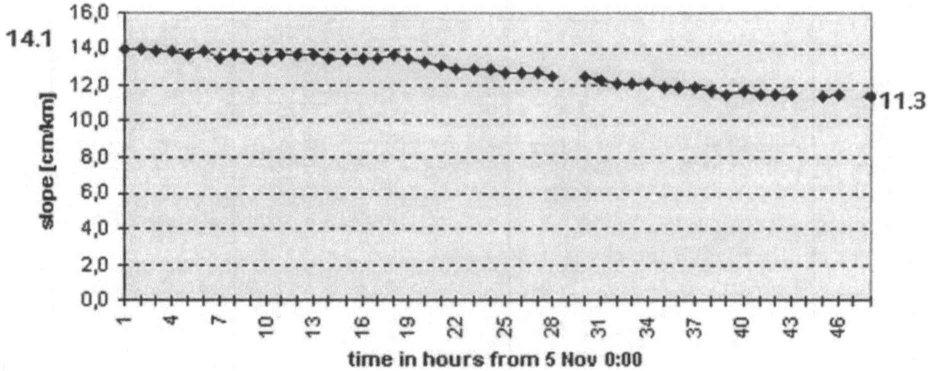

Figure 9. The slope between gauge 862.18 and 867.22 km in the Rhine river, from 5 Nov. 0:00 to 6 Nov 23:00, as the flood wave passes by.

10. Conclusions

Laser altimetry can be used to measure water levels. Comparing the averaged water levels measured with the available gauging stations one can conclude that a considerable accuracy can be achieved using laser altimetry. Moreover it can give water level and slope information — of the water table — of considerable areas whereas gauging stations can not but give local information. Local information on water table may be influenced by obstructions or accelerations of the water. Especially in wide, curved or variably sloping rivers. Water table information may be vital for a proper river management and the protection against flooding.

Perhaps even more important it is that the technique is flexible and that it can be used almost anywhere whenever extreme flood discharges occur. Only in this way it is possible to gather accurate data with regard to the local discharge capacities of rivers.

Furthermore the amplitude of the difference in water levels is spatially distributed in relation to the important river geometry characteristics. It is assumed that using laser altimetry it is well possible to determine wave amplitudes. This application of the technique is especially interesting for coastal zones at seas and lakes where obtaining accurate and substantial wave data may otherwise be difficult to elaborate.

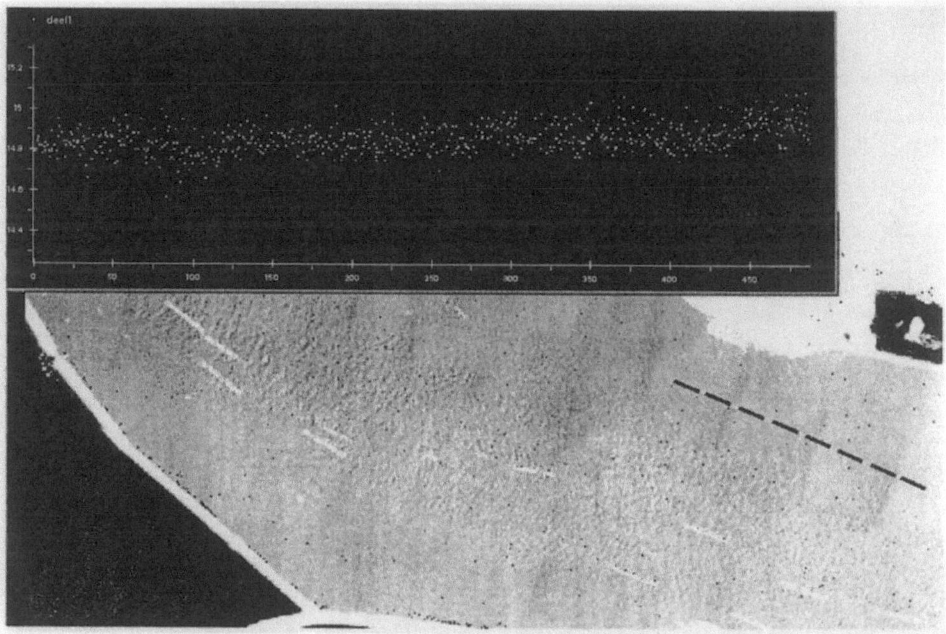

Figure 10. The breakwaters in the Rhine are clearly visible by laser. In the profile the laser points are plotted, along the line over the breakwaters. One can see different wave patterns in the river.

11. References

Huising, E.J. and L.M. (1998) Gomes Pereira, Errors and accuracy estimates of laser data acquired by various laser scanning systems for topographic applications, ISPRS Journal of Photogrammetry & Remote Sensing 53 p. 245-261.

Wouters, W.J.C and A.E. Bollweg (1998) AHN: Actual Height model of the Netherlands, A detailed elevation model using Airborne Laser Altimetry, GeoInformatics, volime 1, september

Lemmens, M.J.P.M. and E.H.W. (6 febr. 1997) Fortuin, Fouten-Analyse van Vliegtuig-Laseraltimetrie (in Dutch), Survey Department Rijkswaterstaat

6. Soil moisture conditions in The Netherlands during the summer of 1995 interpreted from satellite measurements

WIM BASTIAANSSEN[1], HENK PELGRUM[2],
GERBERT ROERINK[2] *and* KAREL SOETERIK[3]

[1] *International Institute for Aerospace Survey and Earth Sciences (ITC), P.O. Box 6, 7500AA Enschede, The Netherlands; AgroWater Consultants, Garststraat 23, 4021 AB, Maurik, The Netherlands*

[2] *DLO-Winand Staring Centre for Integrated Land, Soil and Water Research, P.O. Box 25, 6700AC Wageningen, The Netherlands*

[3] *Technical University Delft, Department of Civil Engineering, Department of Land and Water Management, Delft, The Netherlands*

Abstract

Knowledge on actual evaporation and soil moisture conditions of crops, forests and natural vegetation is of paramount importance for initiating and evaluating water management practices. Despite the scientific progress in hydrology and soil physics, deficit and excessive soil moisture in the root zone cannot be measured or calculated in a straightforward manner due to spatial heterogeneity in environmental conditions. A remote sensing study of the surface energy balance using Landsat Thematic Mapper images has been applied to circumvent the need to solve the water balance explicitly for estimating actual evaporation and soil moisture conditions. The results for various land use categories in The Netherlands are evaluated. It is demonstrated that remote sensing techniques have progressed and that the limited awareness and services in data transfer now restraints the exploration of satellite data for operational water management.

1. Introduction

The Netherlands are wet and flat; the groundwater table is near to the surface in large parts of the country. The long term water balance is provided in Table 1, which reveals that evacuation of excess water from precipitation and inflow from inter-border rivers (total is 85.4 km^3) is crucial to control the groundwater table. Table 1 is based on average numbers, and can be misleading from the perspective of soil moisture availability, because elongated dry spells frequently occur in summers when the crops are fully developed. The maximum daytime air temperature can go up to 25 to 35°C in summer, which increases the unstressed transpiration from vegetation to 5 mm/day

A. van Dijk and M. G. Bos (eds.), GIS and Remote Sensing Techniques in Land- and Water-management, 69–86.

or more. Crop growth will be affected by drought induced stomatal closures, if farmers don't intervene by means of sprinkler irrigation and water boards by keeping the water level in surface drains artificially high. Most arable and horticultural farmers, or a combination of neighbouring farmers, have invested in mobile sprinkler systems, which can be used to prevent crop development stagnation due to moisture deficit. This supplementary irrigation augments the delay in crop growth and enhances crop evaporation. On the opposite, sub-surface drains are installed to enhance the removal of soil moisture excess in spring (sowing/planting season) and autumn (harvest season). However, the central question is whether these techniques and interventions perform adequately.

Table 1. Long term water balance of The Netherlands (gross area 40,844 km^2)

	Volumes (km^3)
Precipitation	30
Inflow from rivers	80
Evaporation from soil	5.9
Transpiration from vegetation	12.3
Transpiration due to irrigation	1.4
Domestic and industries excluding return flow	5.0
Outflow to sea	85.4

Market prices of crops are usually higher during dry summers, thus crop water deficit should be prevented during dry years to obtain maximum yield. The income of arable farmers for several years can be earned in one dry year when crop yield is high, in combination with a good price. Dry summers with less clouds and ample solar radiation for photosynthesis will boost the maximum CO_2 assimilation and crop growth. An interesting experiment between farmers and a national NGO *Centrum voor Landbouw en Milieu* (CLM) was designed in 1996 to develop a practical tool that can help farmers in their decision to sprinkle (*beregeningsplanner*). This manual irrigation planner describes the moment necessary to sprinkle and how much additional water should be supplied to crops or meadows. After two years of experiences on eight selected farms, however, it is concluded that the experiment did not become a success story (Hoving et al., 1998). The major bottleneck is a proper estimation of the capillary contribution of groundwater to the root zone. The applied approach under-estimates the reduction of moisture in the root zone and consequently over-estimates the actual evaporation of the crops.

Although this initiative has created more awareness on soil moisture-evaporation-crop yield interactions in the farmers community, it failed to solve the water balance through the concept of simple calculation methodology. The technical obstacles are:

(i) water flow in soil is a strong non-linear process which can by no means be solved with oversimplified approximation procedures

(ii) local scale variations in the water balance are significant due to heterogeneity of the soil hydraulic properties.

It is next to impossible to simply estimate field scale water balances from general information sources such as a national soil texture map with generalised soil physical properties, generalised crop coefficients and regionally tabulated reference evaporation. Intensive field measurements can solve the problem of data shortage at the local scale, but the regional soil moisture conditions remains unresolved. Hence conceptually new techniques to estimate actual evaporation, crop water stress and soil moisture conditions at a range of scales between fields and regions needs to be developed.

Thermal remote sensing is an attractive solution, because it measures directly the *net effect* that various hydrological and soil-physical processes have on the stomatal aperture: the canopy temperature. The idea of sensing the land surface temperature remotely is almost 30 years old, and origins from Arizona (e.g. Idso et al., 1969; Hiler and Clark, 1971). When a crop is growing under ideal moisture conditions, the canopy temperature is relatively low and similar to the temperature of the overlaying air mass, or to the surface temperature of a nearby lake. If the stomata close for whatever reason, the canopy temperature goes up immediately. Such an easy indicator is suitable for farmers to establish their timing of sprinkling.

Another issue in water resources management of The Netherlands is the public discussion on water consumption by agriculture vs. water used by natural ecosystems and forests. Natural vegetation and less productive areas have been replaced by agriculture during the forties to the sixties. This was a period of food demand and in many cases the water table has declined during that period. As a consequence of a lower groundwater table, it is estimated that 630,000 ha natural vegetation is exposed to man-induced drought. The growing population of the eighties and nineties is now demanding natural ecosystems for leisure and recreational purposes. The public awareness of sustaining historical environments and landscapes is ranking high. This adds pressure to replace agriculture by natural vegetation and naturally low laying brook valley systems. Nature has to be managed also and the question arises how much water will natural ecosystems consume and which soil moisture conditions are favourable for the various species and ecosystems? These questions are very essential to be answered when dealing with competing sectors for rural water use. As water resources are becoming scarcer in the period between June to August, the distribution of water to meadows, wetlands, heaths and peat areas needs to be surveyed together with irrigation requirements for arable land and productive pastures. Adequate regional scale information on these rural water users and their soil moisture conditions can take away myths, and form a sound basis to support the water allocation decision process of water boards and provincial organisations.

Sensing large areas with high resolution spectral radiometers onboard satellites is an option to overcome the problem of determining the water status in agriculture vs. natural ecosystems. As pixels of high resolution satellite images are as small as 30 m, variations in surface conditions within individual farms, forests and ecosystems up to provincial level and beyond can be detected. Radiometers measure reflected and emitted spectral radiances in specific electromagnetic wavelength intervals.

Evaporation algorithms driven by remote sensing data have been developed which can transfer these spectral radiances into maps of actual evaporation and soil moisture status. Several remote sensing evaporation studies have been carried out in The Netherlands on arable land (Soer, 1980; Nieuwenhuis et al., 1985; Thunnissen and Nieuwenhuis, 1989). These studies were, without exception, applied to small agricultural areas and were based on expensive airborne remote sensing. The current study which is also described in Soeterik (1998), is the first endeavour to monitor the soil moisture conditions with space borne multi-spectral data covering large areas in The Netherlands.

The first objective of this paper is to demonstrate that we can technically monitor the evaporation at country level and evaluate the soil moisture status without involving difficult water balance terms and asserted assumptions related to soil moisture flow. The second objective is to re-examine the evaporation figures for different land use categories, which solves myths and lacunas related to water consumption of crops, forests, pastures and natural vegetation. The paper will close with some food for though on how we can exploit the advantages of these advanced information systems better.

2. Material and methods

Landsat Thematic Mapper's visible and infrared spectral radiances of the Central and Southern part of The Netherlands at May 24, July 11 and August 12, 1995, have been used for analysing the actual evaporation (Soeterik, 1998). Each image consists of recordings in seven different wavelength intervals, tree recordings in the visible, three in the near and mid infrared, and one band in the thermal infrared range. The spatial resolution is 30 m except for the thermal infrared band which measures with a 120 m resolution. Landsat 5 has a return period of 16 days.

The Surface Energy Balance Algorithm for Land (SEBAL) is a remote sensing flux algorithm that solves the surface energy balance on an instantaneous time scale and for every pixel of the satellite image (Bastiaanssen et al., 1998). The method is based on the computation of surface albedo, surface temperature and vegetation index from multi spectral satellite data. The surface albedo is calculated from Thematic Mapper bands 1, 2, 3, 4, 5, and 7, the surface temperature from band 6 and the vegetation index from bands 3 and 4. The spatial variability of surface albedo is used to calculate net short wave radiation, and surface temperature for the calculation of net long wave radiation, soil heat flux and sensible heat flux. The vegetation index governs the soil heat flux by incorporating light interception effect by canopies, and is used to express spatial variability in aerodynamic roughness of the landscape. The latent heat flux is computed as the residue of the surface energy balance. Air humidity measurements are not needed because evaporation is computed from the latent heat flux. The SEBAL algorithm has been applied for water balance estimations (e.g. Pelgrum and Bastiaanssen, 1996), irrigation performance assessment studies (e.g. Roerink et al., 1997) and for weather prediction studies (e.g. van de Hurk et al., 1997).

The availability of the complete surface energy balance for each pixel allows the computation of the energy partitioning. If the soil is wet, most of the net radiation is consumed by the latent heat flux, a little to soil heat flux with almost no energy left for sensible heat flux. Under dry conditions, however, the opposite holds true, because latent heat flux will be small, leaving all the energy available to heat soil and air. A way to express this partitioning of the radiant energy, which is gaining international impetus, is the evaporative fraction (e.g. Brutsaert and Sugita, 1992; Crago, 1996). The evaporative fraction is defined as the latent heat flux/net available energy ratio, and a proper indicator of soil moisture available to the crop because the partitioning of energy is controlled by root water uptake from all layers where roots are active. The available energy is the difference between net radiation and soil heat flux. As net radiation, soil heat flux, sensible heat and latent heat flux are all estimated explicitly, a raster map of evaporative fraction can be made. This approach should be considered as a rather new method to express crop water stress and thus the soil moisture status of vegetation for heterogeneous landscapes (e.g. Smith et al., 1992; Bastiaanssen et al., 1997).

Meteorological and hydrological *in situ* data were used to validate the SEBAL estimations of actual evaporation and evaporative fraction for 1998. Due to the present renewed interest in forest hydrology (e.g. Dolman and Kabat, 1993), DLO-Winand Staring Centre for Integrated Land, Soil and Water Research and the Royal Netherlands Meteorological Institute (KNMI) have equipped several forest locations with advanced instruments such as eddy-correlation devices to measure evaporation and other terms of the energy balance (Bosveld, 1997). Among the extensive list of parameters measured in these field experiments are surface solar radiation, surface albedo and surface temperature. The station data used within the present study are two conferious forest sites (*Garderen* and *Kootwijk*), one deciduous forest location (*Zeewolde*) and a pasture location (*Cabauw*), see Figure 1. These ground measurements have been applied to correct the satellite measurements of surface albedo, surface temperature and vegetation index for atmospheric interferences. Meteorological measurements cover by definition only a relative small area, which varies with the height of installation of the instruments above the ground. The evaporation of a 57,000 ha catchment of *Beerze* and *Reusel* in the Southern Province of *Brabant* has, therefore, been included in the validation analysis to assess the performance of the remote sensing methodology for larger regions (Van der Bolt et al., 1996). The evaporation has been drawn from a calibrated regional scale physically based distributed hydrological model (Querner, 1997).

3. Results

The SEBAL-based actual evaporation and evaporative fraction maps have been compared with the fluxes measured in the field and the hydrological simulations of evaporation in *Beerze-Reusel*. Figure 2 shows the error of the SEBAL estimates as a function of areal size. The smallest scale is 2 ha and the error is about 15 to 20%. This area is the approximated footprint of an individual flux tower in either at the grass or at one of the forest sites. The areal integral of the 4 flux sites was also calculated and hypothetically schematised to form one contiguous area of 8 ha. Taking the pixels associated to each flux site and integrate them for a large virtual site of 8 ha, the SEBAL error reduces to 12%. The catchment of 57,000 ha is a composite of the land use units

Figure 1. Location of the flux-sites and the Beerze-Reusel sub-catchment for validating the SEBAL remote sensing algorithm.

agricultural area, woodlands, nature reserves, build up area and surface water, and each sub-unit has a different acreage. The SEBAL comparison was made for each land unit as well as for the entire catchment. The agreement for agricultural areas, natural resources and surface water was quite good, but large discrepancies between the build-up area and the woodlands were found. It is at this stage not clear whether this difference is related to the inaccuracy of the hydrological model or the remote sensing algorithm, but Figure 2 reveals that SEBAL is reasonable accurate for the forests in *Kootwijk*, *Garderen* and *Zeewolde* which suggests that the model may be erroneous for these areas. The error for the total 57,000 ha is 2% only, which shows that the assumptions and default values in SEBAL are acceptable.

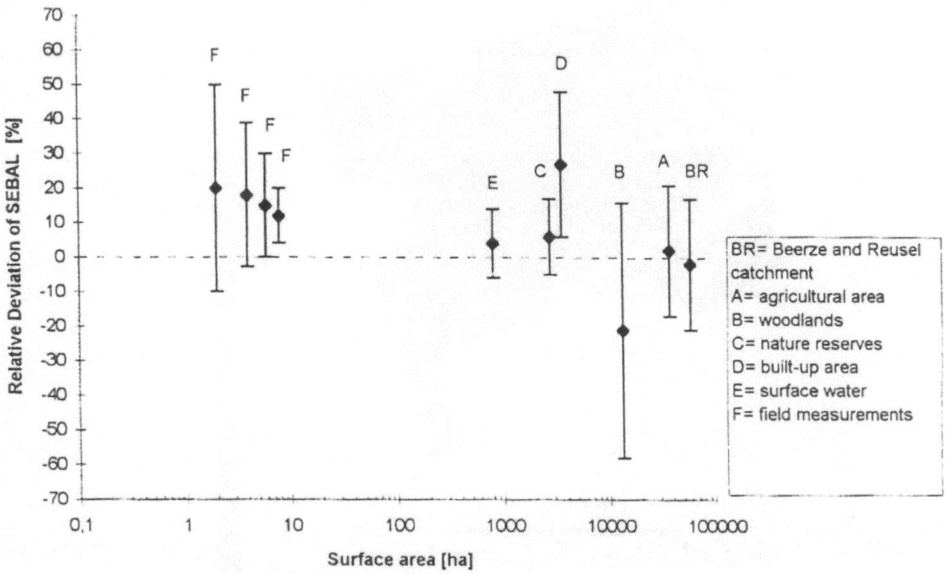

Figure 2. SEBAL error variations of the 24-h actual evaporation with spatial scale.

Figure 3 depicts the results for May 24, 1995. The white holes in the image represent clouds which have been removed from the image. Figure 3 shows that clay soils, which are predominantly cultivated with arable land and meadows for cattle grazing, have an evaporation of approximately 4 to 5 mm/d. The higher pleistocene sandy soils of the provinces *Brabant* and *Gelderland* respond by a significantly lower evaporation at a rate of approximately 2 to 3 mm/d. The open water evaporation of the *Loosdrechtse Plassen* has evaporation rates of approximately 6 mm/d. The detailed inset clearly demonstrate the effect of soil type, depth to the groundwater table and land cover on the actual evaporation: forests on the sandy *Utrechtse heuvelrug* with a water table more than 20 m deep are considerable drier than the surrounding grass lands in the valley of the rivers *Vecht* and *Eem*. The forests in the polder *South Flevoland*, which are expected to be exposed to seepage from the neighbouring inland lake surrounding the polder, exhibit, like the forest on the *Utrechtse heuvelrug*, a strongly reduced evaporation. These forests are thus, for whatever reason, suffering from inadequate moisture conditions.

Drecht and Schepers (1998) and Meinardi et al (1998) published a map prepared at the National Institute of Public Health and Environmental Protection (RIVM) of the longer term actual evaporation for the period 1961-1990 (Figure 4). They have calculated the actual evaporation as the difference between potential evaporation and the evaporation deficit. The potential evaporation was calculated as the product of reference evaporation and a crop coefficient k_c. Following Feddes (1987), the crop coefficient of uncovered soil during winter and spring was kept low (cereals 0.5, legumes 0.4, maize 0.4 potatoes 0.5 etc.). Hence, all arable land in the provinces *Zeeland*, *Flevoland* and *Groningen* with partially fallow land get a low potential

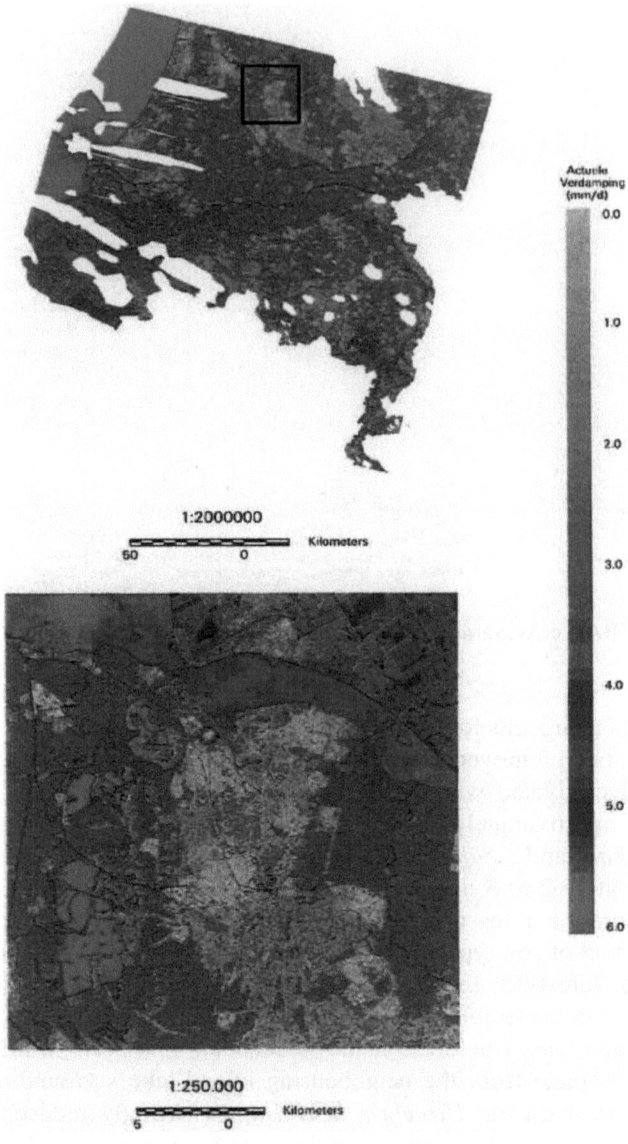

Figure 3. Actual 24-h evaporation of the Central and Southern part of The Netherlands on May 24, 1995, computed from Landsat Thematic Mapper images. The inset depicts the variation of a traditionally low peat landscape, a pleistocene sandy area and a strip of newly reclaimed polder land.

evaporation during their fallow winter season. The evaporation deficit is calculated from the national groundwater depth classes inventory (class I to VII) and the soil data base using simplified hydrological calculations (HELP Table, 1987).

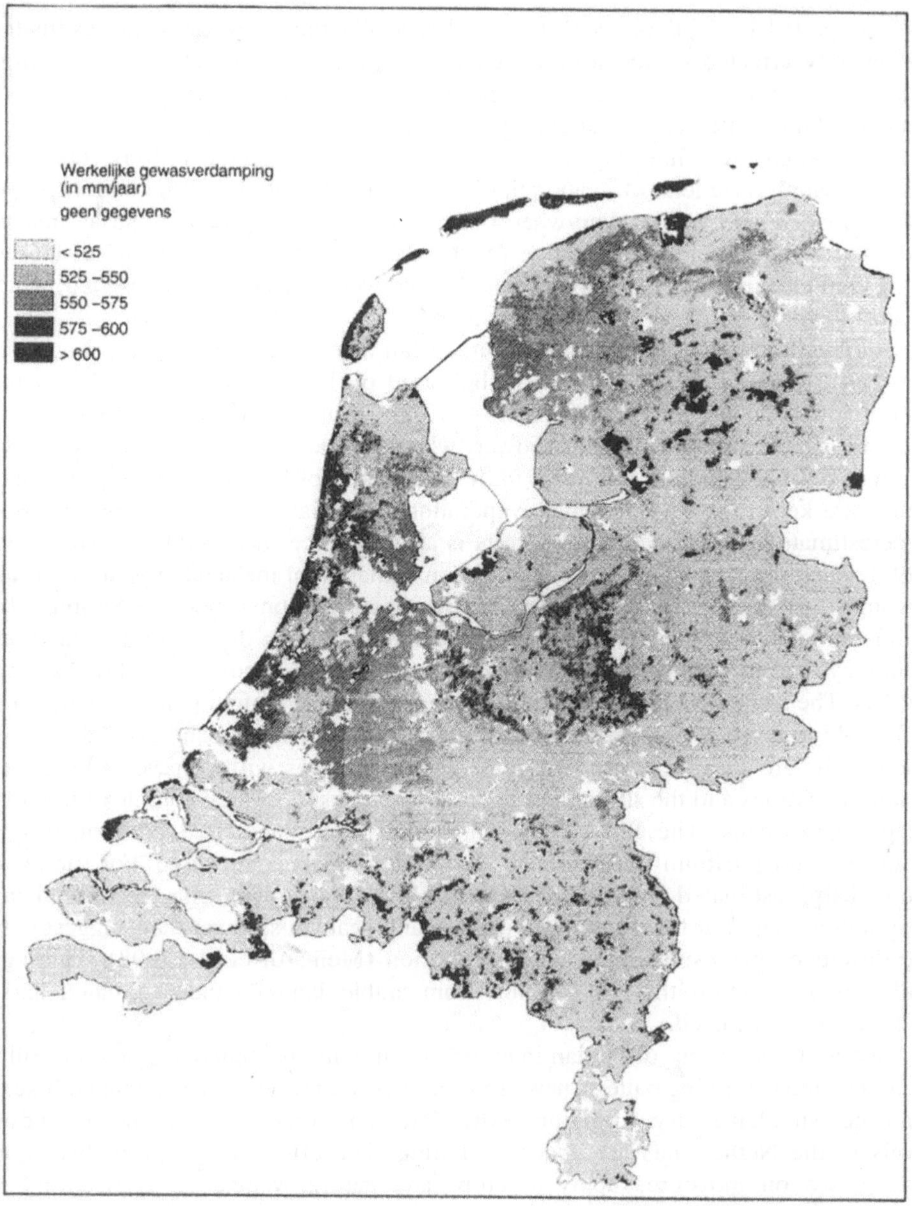

Figure 4. Long term actual evaporation of The Netherlands (1961-1990) estimated with a simplified hydrological calculations in a GIS environment, RIVM, Bilthoven (after Meinardi *et al.*, 1998).

The time scales between Figure 3 and 4 are different, and the two maps should therefore not be compared. But there is too much contradiction to ascribe the differences to the time factor alone. Sub-regions which the RIVM map describes as being wet are interpreted by SEBAL as being dry. Especially the forest areas are estimated to be severely affected by moisture deficit on the SEBAL map, but receive the highest evaporation on the RIVM map. The opposite can be noted for the potatoes and sugar beets in *Flevoland.* The areal patterns of the other 2 Landsat images acquired in July and August, are not very much different from the situation in May (not shown). As the spatial variations of evaporation in wet Dutch winters will be extremely small (it is clouded and moist everywhere), the average of the three Landsat images is expected to be rather representative for the areal patterns of the annual evaporation. So, a kind of overall agreement between dry and wet land should be expected, which is against the patterns observed in Figures 3 and 4.

Although no definite answer on accuracy can be provided, part of the contradiction between Figures 3 and 4 is likely to be based on two strong simplifications in the hydrological calculation procedure. Firstly, the evaporation is near to potential if the bare soil is wet after a rain shower, which frequently occurs in the autumn and summer season. This will not occur in the calculation of k_c values for sparse canopies, which are kept low. The potential evaporation of partially fallow land is thus grossly underestimated. The k_c value for forests is also a source of uncertainty (Moors et al., 1994) which may introduce a systematic overestimation of the actual forest evaporation. Secondly, as mentioned before, the capillary contribution cannot be simplified and estimation techniques or surrogates for unsaturated Darcian fluxes are in their infancy. That may lead to wrong flux estimations toward the root zone and invalid evaporation deficits. The same conclusion was also drawn from the irrigation planner, which makes it feasible to draw a more general conclusion: generic soil data and groundwater table fluctuations are insufficient to determine capillary fluxes with a reasonable accuracy. Spatial variations and the strong non-linear flow processes are too complex for applying simple approaches. The RIVM map on rainfall minus actual evaporation has been validated with a tritium tracer to estimate the groundwater recharge. For sandy soils, the recharge estimated from tracers was in a good agreement with the rainfall minus evaporation map, but for occasions with clay and peat soils, lower than suggested from the difference between rainfall and evaporation (Meinardi et al., 1998). This agrees with the observation that evaporation from arable land is underestimated and the recharge overestimated.

Due to a wide array of human interventions in land use (encroachment of build up areas, changed cropping pattern, new crop varieties) and water management (sub-surface drainage, supplementary irrigation, extractions for domestic use), the groundwater levels in the Netherlands are generally falling. The effect of the groundwater table draw down on moisture supply to crops and natural vegetation is not very well understood. A desiccation map for a mixture of ecosystems was made by the Ministry of Traffic, Public Works and Water Management (Rijkswaterstaat, 1989). The class on the desiccation map being identified as "strongly affected" is located in the *Utrechtse Heuvelrug*, the *Peel* and the North Sea dune stretch. A strong correlation exists with the map of evaporative fraction presented in Figure 5. The triggering of water stress in vegetation is biome dependent, but generally a reduction of the potential evaporation starts when the evaporative fraction falls below 0.75 to 0.85. Figure 5 demonstrates that all pleistocene sands have an evaporative fraction of 0.1 to 0.5 by the end of May, which implies a severe water shortage and increased temperatures and oxygen concentrations, promoting nitrogen mineralisation in wet land eco-systems. This example demonstrates, the mapping of drought prone areas can be enhanced through the inclusion of spatially distributed satellite data.

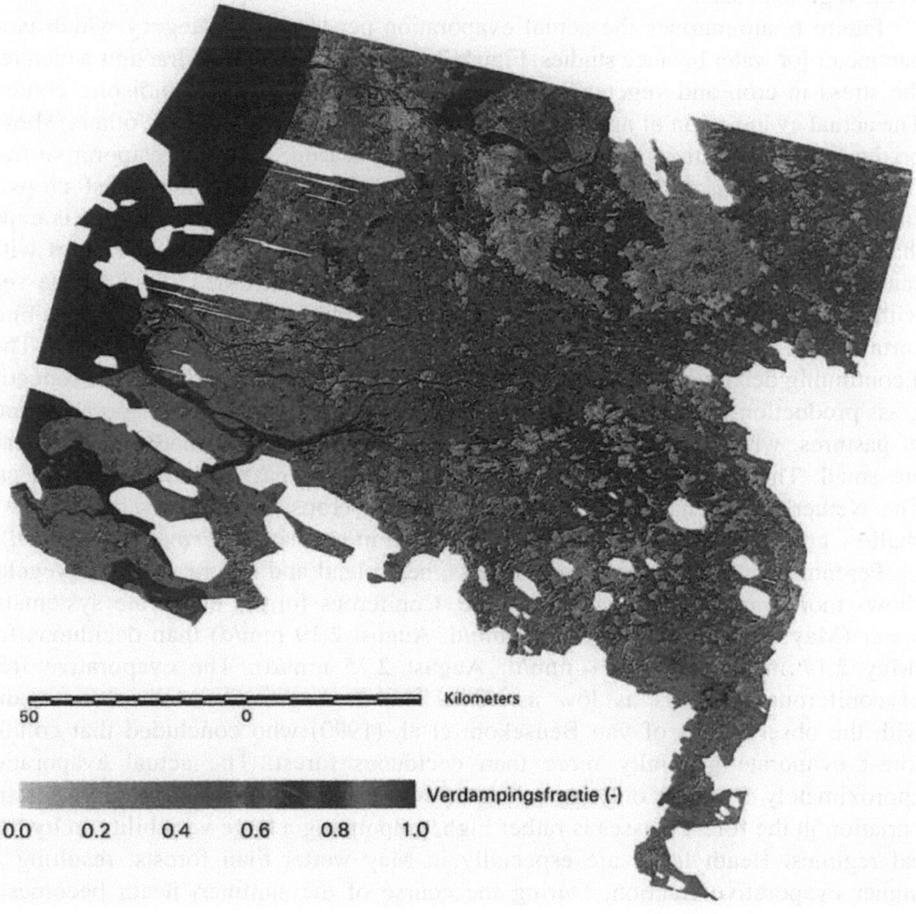

Figure 5. Evaporative fraction of the Central and Southern part of The Netherlands on May 24, 1995, computed from Landsat Thematic Mapper images with the SEBAL algorithm.

4. Land cover and evaporation

Thunnissen and Noordman (1996) compiled the National Land Use database (LGN) from Landsat Thematic Mapper images, complemented with field visits and topographical and cadastral information. The LGN version 1995 was based on the same Thematic Mapper images as used for the current study. Combining LGN with the evaporation maps provides a unique opportunity to get a general overview on the evaporation and soil moisture status of The Netherlands for each of the 25 land use categories discerned in the LGN legend. This goes a step further than local experiments and modelling efforts on crops and forest evaporation, and can be used to extrapolate local studies to the regional scale.

Figure 6 summarises the actual evaporation per land use category which is a key parameter for water balance studies. Figure 7 shows the evaporative fraction which reveals the stress in crop and vegetation development due to inadequate moisture conditions. The actual evaporation of arable crops (maize, potatoes, beets, cereals, others) show surprisingly little variations throughout the growing season. Also the evaporative fraction for arable crops is rather similar, around 0.8, which implies that most crops have sufficient access to soil moisture. Although not systematically analysed, it is expected that the lower bound of the evaporation and evaporative fractions coincides with the sandy soils and the upper bound with soils having finer textures. This can be verified with overlaying the digital soil map. Sprinkling with groundwater is prohibited in certain areas to arrest the falling water table and to reserve natural eco-systems. There is a continuing debate on how much effect sprinkling of meadows has on the reduction in grass production. The data in Figure 7 illustrates a small spread in evaporative fraction of pastures, which suggest that the variation in soil moisture conditions in meadows are small. The evaporation behaviour of the extensive pastures in the central part of The Netherlands behave the same as the arable crops, most likely due to the very shallow groundwater table of 20 to 50 cm depth in peat soils (*veenweidegebieden*).

Perennial vegetation (orchards, forests, heath land and nature with low vegetation) shows more variation in space and time. Coniferous forests evaporate systematically lower (May: 1.76 mm/d; July 2.56 mm/d: August 2.19 mm/d) than deciduous forests (May 2.17 mm/d: July 3.01 mm/d: August 2.75 mm/d). The evaporative fraction of coniferous forest is as low as 40%. This finding is in a direct contradiction with the observations of van Beusekom et al. (1990) who concluded that coniferous forest evaporate generally more than deciduous forest. The actual evaporation is approximately 2 mm/d only, and half of what arable crops consume. The standard variation in the forest classes is rather high, pinpointing a huge variability in hydrological regimes. Heath lands are especially in May wetter than forests, resulting in an higher evaporative fraction. During the course of the summer, heath becomes drier than deciduous forest. The LGN class name 'nature with low vegetation' consists of herbaceous vegetation and turns out to form the class of wettest natural vegetation, although being systematically drier than arable crops and forests. It can be found on the *Veluwe*.

The highest evaporation occurs from the open water surfaces with 5 to 7 mm/d. The evaporative fraction of fresh and salt water lies very near to one. Seawater evaporation is higher than of inland waters due to more available energy; the sensible heat flux of the deep salt water is lower due to a lower water temperature than the water temperature of the fresh inland lakes with shallow depth. The lowest evaporation occurs in The Netherlands in the LGN class continuous urban area, but this mapping unit can still evaporate 2 to 3 mm/d, due to city parks, lane trees and many privately owned and sprinkled small lawns.

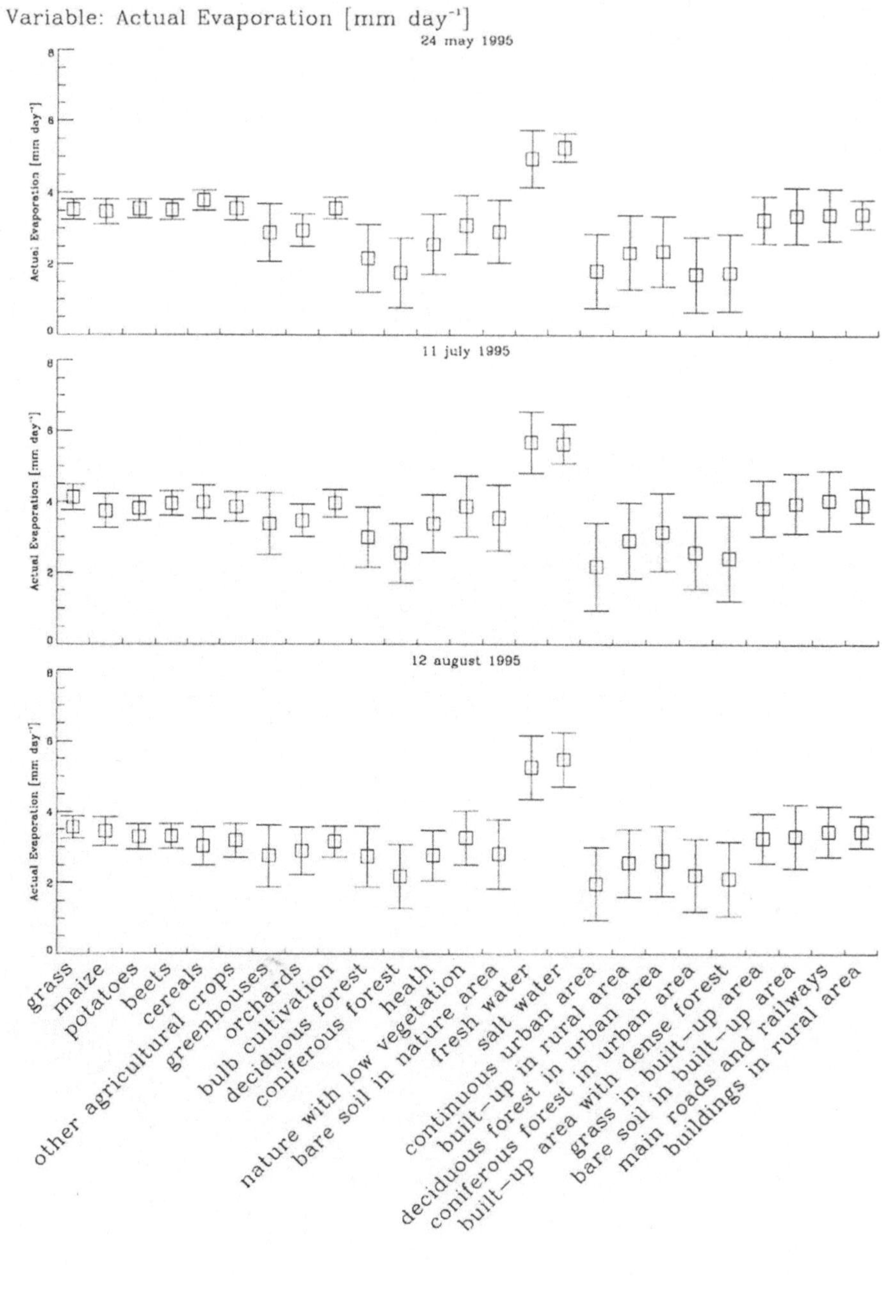

Figure 6. Mean and standard deviation of actual evaporation during the summer of 1995 of land use classes classified according to the National Land Use database (LGN).

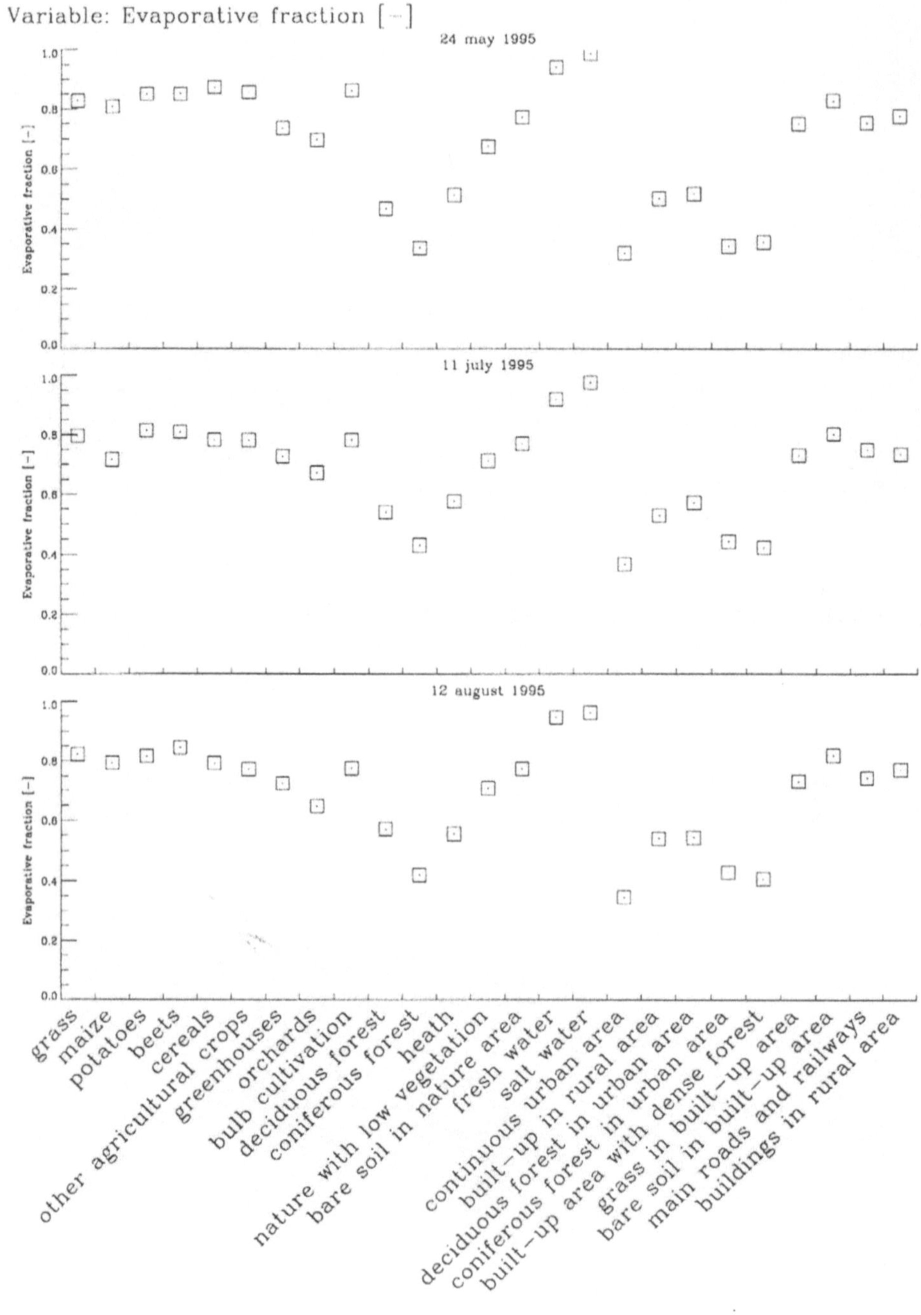

Figure 7. Mean and standard deviation of evaporative fraction during the summer of 1995 of land use classes classified according to the National Land Use database (LGN).

5. Conclusions and future directions

5.1. Achievements

Water management is essential for agricultural production and sustaining environmental rich ecosystems, even in relatively wet countries such as The Netherlands. The soil water balance is the classical basis to evaluate the sufficiency of water supply to the roots of vegetation. Sub-surface soil moisture flow can, however, not be straightforwardly estimated or modelled unless extra-ordinary field and laboratory measurements are conducted. This hampers the description of the needs to irrigate crops, rise groundwater tables in naturally vegetated areas or to assess drought damage due to falling water tables at the regional scale. It is concluded that the actual evaporation and stress due to moisture deficit for the different land use categories at national level is only prematurely known, despite manifold efforts in agro-hydrological and soil-physical research.

The advantage of remote sensing techniques is its capability to measure directly the physical vegetation conditions (temperature, albedo, vegetative cover) being the net result of various sub-surface moisture flow processes. There is no need to describe the sub-surface physical processes if evaporation can be explicitly calculated from remote sensing data and energy balance algorithms. The interpretation of remote sensing data into evaporation has a physical basis and can be applied to other agro-ecosystems. The spatial variability of the parameters in the SEBAL model are described in an empirical manner, but Figure 2 has demonstrated that the accuracy is sufficient to describe the processes at 10 ha and for larger regions (error smaller than10%). Arable crops and grass land have similar actual evaporation rates. Deciduous and coniferous trees evaporate in dry summers significantly less than agricultural land. Coniferous forest consume consistently less water throughout than deciduous trees during the summer season. The spatial variation in evaporation of forests and natural vegetation is much larger than for arable crops and meadows.

5.2. The provider

The remote sensing technique has advanced during the last 30 years (Bastiaanssen, 1998). It has started off as a promising technique after the launch of the first Landsat satellite in 1972. The high expectations of remote sensing technologies were never realised due to many complex physical interactions between the remotely measured signals and the final parameters to be determined at the land surface. However, the scientific achievements have gradually progressed in certain fields (although manifold unresolved issues exist), and the time has ripened to open the dialogue with potential end-users of remote sensing data. The clients are deservedly critical, so the transfer of remote sensing technology to support water management should go slowly and together with demonstration studies. The costs of remote sensing data and interpretation has come down dramatically by the reduction of software and hardware prices in the world of information technology. A personal computer with simple and cheap raster and vector map software has nowadays sufficient capacity to analyse complex Landsat images.

Images from NOAA-AVHRR with a spatial resolution of 1.1 km and a return period of one day are now freely available on internet. The newly launched Landsat-7 has an Enhanced Thematic Mapper and will sell full scenes at a rate of $ 600, being a factor 10 less than the current Landsat-5 Thematic Mapper images. This all implies that the costs related to operationalization of remote sensing technologies is nowadays essentially governed by the wages of the technicians performing the analyses. It should be recognised that application software dedicated to hydrological and water management applications is under-developed, which affects the speed of dedicated image processing negatively.

Access to hydrological data based on remote sensing measurements is neither common in The Netherlands, nor abroad. The number of consultants carrying out commercial remote sensing analysis is fortunately growing, which may open the door to market these products. As many farmers have access to internet — the web sites depicting radar images on cloud cover and precipitation are frequently visited by farmers — they can technically study the soil moisture conditions on their own fields, if a provider opens a web site related to this topic. With windowing and zooming options, farmers can use internet to study the soil moisture conditions at their own farm. The same possibilities applies to water boards or foresters. Hence, technically, the information does not need to go through lengthy hierarchical levels in the governmental sector, but can go directly from the provider to an end-user. There is however no national provider on soil moisture conditions in the rural areas, simply because there is no client and the potential end-user does not know that the technique is ready. This vicious cycle needs to be broken by fostering demonstration studies. The Netherlandes Remote Sensing Board (BCRS) helps financially to boost demonstration projects, but this program closes down in 2000.

5.3. The customers

Water boards and farmers could be customers of remote sensing data. This community has demonstrated to have adaptive skill to apply high tech solutions (automatic weather stations, automatic night frost prevention systems, automatic water distribution systems, climate control in stored products, precision farming etc.). Farmers are economically punished if they don't manage their resources well. Farmers with economically valuable crops such as flower bulbs, vegetables and seed potatoes will be interested to know the growing conditions on their farms.

Other customers are the provincial organizations. Large parts of the pleistocene sands are exposed to human-induced drought with deteriorating habitat conditions for natural vegetation. The 12 provinces have their own methodology to map water deficit conditions which induces step changes at the provincial borders. Figures 3 and 5 show how quantitative assessments can be made at a range of scales between 30 m to provincial level. This could be highly useful to obtain a thorough understanding of drought at the regional scale for provincial organisations. Also national water related institutions such as the Inland Water Management and Waste Water Treatment (RIZA), National Institute of Public Health and Environmental Protection (RIVM) and domestic water supply organisations and water boards involved in regional hydrological studies are possible users of remotely sensed hydrological information.

It seems mandatory to create interest and enhance the public relations if remote sensing will ever be regarded as a professional tool. The initiative to establish operationalization of remote sensing technology lies primary at the side of the Dutch remote sensing scientific community. If they fail to send out opportunistic messages describing successfully tested methods, the technique will remain a research tool forever.

6. References

Bastiaanssen, W.G.M., H. Pelgrum, P. Droogers, H.A.R. de Bruin and M. Menenti (1997) Area average estimates of evaporation, wetness indicators and top soil moisture during two golden days in EFEDA, Agr. and Forest Met. 87: 119-137

Bastiaanssen, W.G.M., M. Menenti, R.A. Feddes and A.A.M. Holtslag (1998) A remote sensing surface energy balance algorithm for land (SEBAL), part 1: formulation, J. Of Hydr. 212-213: 198-212

Bastiaanssen, W.G.M. (1988) Remote sensing in water resources management: the state of the art, International Water Management Institute, Colombo, Sri Lanka: 118 pp.

Beusekom, van, C.F., J.M.J. Farjon, F. Foekema, B. Lammers, J.G. Molenaar and W.P.C. Zeeman (1990) Guide for groundwater management for nature, forest and landscape (in Dutch), Sdu Uitgeverij, 's-Gravenhage

Bolt, F.J.E., van der, P.E.V. Walsum and P. Groenendijk (1996) Nutrient deposition on surface and groundwater in the catchments of the Beerze, Reusel and Rosep. Simulation of the regional hydrology (in Dutch), Rapport 306, DLO-Winand Staring Centre, Wageningen, The Netherlands

Bosveld, F.C. (1997) Derviation of fluxes from profiles over a modern homogeneous forest, Boundary Layer Meteorology 84: 289-327

Brutsaert, W. and M. Sugita (1992) Application of self-preservation in the diurnal evolution of the surface energy budget to determine daily evaporation, J. of Geophysical Res., vol. 97, D17: 18, 377-18.382

Crago, R.D. (1996) Conservation and variability of the evaporative fraction during the daytime, J. of Hydr. 180: 173-194

Dolman, A.J. and P. Kabat (1993) Drougth and water management of forests (in Dutch), Bosbouw tijdschrift: 119-122

Drecht, G. and E. Schepers (1998) Updating model NLOAD for nitrate leaching of agricultural soils: description of model and GIS environment (in Dutch), RIVM rapport 711501002, Bilthoven

Feddes, R.A. (1987) Crop factors in relation to Makkink reference crop evaporation, in (ed.) J.C. Hooghart, Evaporation and Weather, TNO-CHO proceedings and information no. 39: 33-45

Hiler, E.A. and R.N. Clark (1971) Stress day index to characterize effects of water stress on crop yields, Transactions of the ASAE 14: 757-761

Hovink, I.E., H. Everts and J. Alblas (1998) Customised sprinkle irrigation 1997, trialling of the irrigation planner and irrigation guide (in Dutch), Rapport 172, Praktijkonderzoek Rundvee, Schapen en Paarden, The Netherlands: 24 pp.

Hurk, van den, B.J.J.M, W.G.M. Bastiaanssen, H. Pelgrum and E. van der Meygaarden (1997) Soil moisture assimilation for numerical weather prediction using evaporative fraction from remote sensing, J. of Applied Met. 36: 1271-1283

Idso, S.B., R.D. Jackson, W.L. Ehrler and S.T. Mitchell (1969) A method for determination of infrared emittance of leaves, Ecology 50: 899-902

Meinardi, C.R., C.G.J. Schotten and J.J. de Vries (1998) Groundwater recharge and surface runoff in The Netherlands, longer term averages for sand and loamy soils (in Dutch), Stromingen 4 (3): 27-41

Moors, E.J., A.J. Dolman, W. Bouten and A.W.L. Veen (1994) The evaporation of forests (in Dutch), Internal Note 323, DLO-Winand Staring Centre, Wageningen, The Netherlands: 13 pp. Rijkswaterstaat (1989. Ministerie van Verkeer en Waterstaat, Water for now and later, (in Dutch) 3[e] Nota Waterhuishouding, SDU Uitgeverij, 's Gravenhage, 21 250, nrs. 1-2: 297 pp.

Nieuwenhuis, G.J.A., E.M. Smidt and H.A.M. Thunnissen (1985) Estimation of regional evaporation of arable crops from thermal images, Int. J. Of Rem. Sens. 6(8): 1319-1334

Pelgrum, H. And W.G.M. Bastiaanssen (1996) An intercomparison of techniques to determine the area-averaged latent heat flux from individual *in situ* observations: a remote sensing approach using EFEDA data, Water Resources Research 32(9): 2775-2786

Querner, E.P. (1997) Description and application of the combined surface and groundwater flow model MOGROW, J. Of Hydr. 192: 158-188

Roerink, G.J., W.G.M. Bastiaanssen, J. Chambouleyron and M. Menenti (1997) Relating crop water consumption to irrigation water supply by remote sensing, Water Resources Management 11(6): 445-465

Smith, E.A., A.Y. Hsu, W.L. Crosson, R.T. Field, L.J. Fritschen, R.J. Gurney, E.T. Kanemasu, W.P. Kustas, D. Nie, W.J. Shuttleworth, J.B. Stewart, S.B. Verman, H.L. Weaver and M.L. Wesely (1992) Area-averaged surface fluxes and their time-space variability over the FIFE experimental domain, J. of Geophysical Research, vol. 97, no. D17: 18,599-18,622

Soer, G.J.R. (1980) Estimation of regional evaporation and soil moisture conditions using remotely sensed crop surface temperatures, Rem. Sens. Env. 9: 27-45

Soeterik, K.L. (1998) Estimating of seasonal evaporation and water stress in the Netherlands using Landsat images, M.Sc. Thesis, Technical University Delft, Department of Civil Engineering,

Section Land and Water Management: 75 pp. HELP Table (1987) Influence of water management on agricultural production, (in Dutch), Rapport van de Werkgroep HELP-Tabel, Mededeling Landinrichtingsdienst no. 176

Thunnissen, H.A.M. and E. Noordman (1996) Classification methodology and operational implementation of the land cover database of the Netherlands, Report 124, DLO-Winand Staring Centre, Wageningen, The Netherlands

Thunnissen, H.A.M. and G.J.A. Nieuwenhuis (1989) A simplified method to estimate regional 24-h evaporation from thermal infrared data, Rem. Sens. Of Env. 31: 211-225

7. GIS and watermanagement: Where are we heading to?

ERIC J. VAN CAPELLEVEEN *and* FRED H.W.M. VAN DEN BOSCH
Twynstra Gudde Management Consultants, PO Box 907, 3800 AX Amersfoort, The Netherlands
<ECA@TG.NL>

Abstract

Both the world of Geo-ICT and Water management are at this moment subject to radical changes. The application of data-acquisition technologies like Remote Sensing, Laser altimetry and GPS-based fieldequipment rapidly grows. The availability of low-key GIS desktop- and pentipsoftware allows users to map Geo-data by themselves. OPEN GIS compliant software, launched in April 1999, will stimulate users to exchange Geo-data. Revision of common water management techniques, introduction of integral watermanagement and combination of watermanagement with spatial planning are coinciding with the current boost in Geo-technology. Finding the correct application of technology to serve both politicians and watermanagement professionals will be the issue in the coming years. Remote sensing will without any doubt prove to be a good and cost effective datacollection tool. Simulation and visualisation tools can present understandable maps providing quick pattern recognition, supporting planning and decisionmaking. Using all these fantastic possibilities of technology may become a pitfall. The bulk of data, analysis and simulation seems to contradict the reduction of complexity which must be strived for. Focusing of the topics, monitoring the attention for predictability, justification of planning and decisionmaking is recommended.

1. Remote Sensing supports integral watermanagement

Remote Sensing is widely known as a technique to collect Geo-data from space and to use this data to investigate in developments on our planet. The dutch beer commercial (Dommelsch) stepwise focusing on the spot, revealing a tempting pint of beer on a terrace, introduced Remote Sensing possibilities to the public. As an input device for management, preventing disliked surprises, Remote Sensing is less known. Yet Remote Sensing, being one of the spatial data collecting techniques, enables us to evaluate natures behaviour quickly and frequently. As mankind has interfered in nature i.e. brooks, rivers, polders and increased the paving ratio, we are now facing the need to re-evaluate our watermanagement. A combination of Remote Sensing and models can provide both technicians, governmental officials and managers a set of tools to plan and monitor the new approach on watermanagement. In this paper we explore on some trends in the field of spatial data and watermanagement. We shall illustrate the expected application of new technology and policy due to the distinguished trends in some examples.

A. van Dijk and M. G. Bos (eds.), GIS and Remote Sensing Techniques in Land- and Water-management, 87–92.

Before making a wrap-up we have enumerated some considerations due to the spotted change, thus providing the generally welcomed context on new technology and new policy.

2. Trends and developments

In spatial information management and water management we have spotted following trends:

water management:

- increasing need and political interest for integral planning and decisionmaking processes (water management becomes a multi linked issue)
- combining governmental tasks concerning watermanagement to improve governmental efficiency and their ready wit
- a huge task renewing sewer systems
- improving water quality (neutralisation Ph-> 7)

spatial information management (GIS):

- improved data collection techniques and improved data accessibility provide a bulk of Geo-data
- web mapping (GIS on the Internet)
- OPEN-GIS interoperability standards and tools
- modelling and interactive decision support.

Let's drill down to this list of trends and developments before explaining the application of Remote Sensing in combination with other spatial management trends for integral water management.

2.1. Trends concerning water management

2.1.1. *Political interest for integral planning and decision making*

Current developments show an increasing interest of policy makers in integral planning and decision making. Water management used to be excluded from this integral planning. The problems we are facing due to coinciding abundant fall of rain and snow resulting in floods and avalanches underline the need for integral planning. Water, and its abundant and sudden appearance, must be taken into account while planning roads, building locations and recreation areas. Emergency issues must be added to the list of integral aspects.

2.1.2. *Need to improve efficiency and ready wit*

Combining governmental tasks concerning water management to improve efficiency and ready wit is a second trend which can be signalled. Water quality, quantity management as well as managing and administrating dikes and sewers are traditionally spread over a series of governmental organisations. Current costs have urged politicians and governmental officials to improve efficiency.

Concentration of nearly all water management tasks to one organisation is the logical outcome of this efficiency drive.

More and more extraordinary circumstances concerning too much and to too less water (floods and drought) urge water managers to improve the readiness and ready wit of their organisations and drainage and buffering systems.

2.1.3. *Renewing sewer systems*

The age of our present sewer system implies a renewal in the next decade. This enables us to re-evaluate the objectives of most sewer systems to transport the water, both rain and waste water, as quick as possible to the rivers and sea. Buffering can be implemented in the new systems and thus a spread of flow can be achieved preventing high flows of water to be handled by rivers.

2.1.4. *Improving water quality (neutralisation)*

Maintaining the quality of water is an increasing issue for producers and distributors of drinking water. Due to intensive cattle, pig- and poultrybreeding, the extraction of groundwater, and the recharge of aquifers with water from polluted rivers, and the process of maintaining and improving the quality of water has become a major task. The need for groundwater management nearly equals the need for surface watermanagement.

2.2. Developments regarding spatial information management

The most appealing developments in the world of GIS are discussed below.

2.2.1. *Bulk of Geo-data*

Improved data collection techniques like photogrammetry, Remote Sensing, laser altimetry, GPS based levelling etc. as well as improved data accessibility on Geo-data files have resulted in a bulk of Geo-data. Handling the bulk of data and filtering it cause a growing importance of spatial datamanagement. The growth of the databulk is stimulated by the still growing size of the infrastructural and urban spatial planning projects.

2.2.2. *Web mapping (GIS on the Internet)*

GIS applications are now available to be run over the Net. The actual applications vary from 'dumb' client applications that enable users to run a GIS or GIS alike program on the server, using the Web for datacommunication, up to the intelligent client applications that download both a part of a application program and data and are able to execute the spatial queries and presentations on the client machine. In both ways GIS functionality is made available to a wide group of users by just using the Web. Boston's MIT[1] has proven that even live panning and zooming of orthographics can be made accessible at good response conditions for remote users using the Net. We predict a boost in the use of these spatial possibilities in all kinds of applications.

[1] Prof. Ferreira. Massachusetts GIS

2.2.3. *OPEN-GIS interoperability standards and tools*

The first OPEN-GIS interoperability standards compliant programs and tools have emerged in April of 1999. They will enable organisations to exchange/access Geo-data from various format using one tool without labour intensive conversions. The traditional hurdles preventing users to combine various sets of Geo- and administrative data will be demolished rapidly. It will however force organisations to label their data (meta data) in order to enable data use by others. Yet, a more frequent data exchange will emerge.

Apart for the Geo-data issue, OPEN-GIS enables application program builders to create a library of program modules which will be able to inter-operate on data packages, thus enabling the user to pick and select application blocks for rapid application development. We expect users to create GIS alike applications more and more by themselves.

2.2.4. *Modelling and interactive decision support*

Making models of natures behaviour trying to predict the consequences of human interference, is used for some time now. Applying 3D presentation, combining GIS and CAD functionalities, time-based snapshots (4D) and tracking facilities showing the dynamic pattern between the snapshots are seen more often. These models which regularly contain more and integral aspects are even represented to its users by interactive groupware decision support systems (GDSS). These GDSS systems enable a group to interactively influence the outcome of a decision process by prioritising the destiny of areas interactively.

3. Water management benefits from Remote Sensing

Let us illustrate where GIS/RS can support decisionquality and/or planning and monitoring regarding water management in some examples.

3.1. Spatial and environmental planning

The political request to integrate water management and urban and spatial planning can be supported by Remote Sensing. This can be done using RS acquired data like changes in groundwater level of an area due to intensive rain to model the effect of slower water transport, buffering and urbanisation in that area. Especially on those locations where parcel reallocation must be combined with the creation of new recreation and environmental protection measures, Remote Sensing acquired Geo-data can provide a quick insight on possibilities and potential bottlenecks. Orthographs can be used to downsize the threshold standard RS rasterbased maps evoke.

Monitoring the effect of urbanisation in watery areas can also be supported by RS obtained data. Planned changes in flow capacities and water storage capacity can be monitored and conclusions can be reached to improve the watermanagement in these areas.

3.2. Renewing sewer systems

The planned renewal of sewer systems, possibly introducing a separate system for rainfall and domestic polluted water (sewer), can be supported by 4D GIS simulating the flow in the system due to input whilst renewing and separation. Remote Sensed data on the pollution of rivers, canals and lakes due to a thermal, chemical or sewer pollution can also provide data to reduce the impact of such pollution in the future. RS data will also enable the monitoring of planned and executed measures regarding the pollution or renewed sewer systems.

3.3. Democracy using the web

Determination and visualisation of for instance the cost effectiveness of water management measures and facilities is more and more required. Looking up decisions, plans and the actual water conditions (rainfall, water quality and groundwater level) will be desired by civilians and enterprises. Web mapping, where maps, showing the desired data, can be retrieved using the Internet and can provide the desired insight. These maps can be either prefabricated maps or interactively produced maps. Anyhow, it would explain which cost are made to manage water in an area and stimulate the openess of water management organisations. Remote Sensing could partially provide the data thus enabling the water management organisation to inform those involved about intended facilities and measures. As where the Dutch Ministry of Public Works and Transport already has announced that the Netherlands will recreate some overrun/flow areas that shall store the abundent volumes if nature blesses us with excessive rain. The civilians and companies involved will appreciate timely information about these plans, their effect and the impact on their daily lives. Predictability, justification and openess are virtues of modern integrated management. Geo-ICT such as RS data can assist the accomplishment of these virtues.

4. Requirements to the sophisticated use of Remote Sensing

The application of Remote Sensing to enable sophisticated analysis and presentation of watermanagement issues will however introduce the need to pay more attention to some management issues. Thus we would emphasise to keep following issues in mind.

4.1. Managing the bulk of data

Attunement of dataquality to its intended use and constant awareness of the needed data-accuracy in regard to the critical bandwidth for decision making from a managerial and political point of view needs special interest. Constant evaluation on internal datacollection or grabbing the data from the Geo-data Internet pile should become a natural capability of both Geo-information professionals and their management. OPEN-GIS standards will facilitate this process.

4.2. Reducing complexity

Multiple space-use and integral policy making will sweep up the need for scenarios. Modelling these scenarios can be performed in a range of scales and areas of interest. An early setting of priorities on possible scenarios will without any doubt shrink the complexity correlated with integral policy making without losing any decision quality or justification. Frequent, open and plain communication towards those involved using the Remote Sensing based maps is required to explain the elimination process.

4.3. Improving efficiency

Next to the need to improve efficiency both in datacollection and analysis, socially it is no longer acceptable tó omit preventative measures even if they imply a cost raise. A pro-active policy to evaluate a What-if analysis using relatively cost-effective remote sensing data shall be required. Improving efficiency from a watermanagement point of view, may imply buffering above transporting. Soil conditions in conjunction with the watertransportsystem capabilities determine the effectiveness of such natural water-management systems. Remote Sensing can assist building and optimising of models by proving the data during a pilot period in a specific area. Apart from the need to improve efficiency and reduce costs, public opinion about the omission of vital measures may cause the need to activate preventive measures.

This causes a social trend to execute pro-active policies and to communicate about its justification. WHAT-IF will be a major issue which needs support. RS can be an essential tool to discover trends or to predict catastrophes.

4.4. Predictability and justification

Predictability and justification are two very important virtues for government. As radical choices with sometimes drastic consequences may be expected, government should be able to (re)produce the origination of its decisions and intended plans. The value of land and property can be highly influenced by the choices government intends to or already has made. These choices will much more than before be dependent of watermanagement issues. Both civilians and enterprises will demand a clear and reproducible decisionmaking course. As maps generated by remote sensed Geo-data will be used to justify the choices, these analysed maps should be made accessible to the public for instance using the webmapping techniques and Geo-data access using OPEN-GIS standards.

4.5. Value for money

To gain value for money implies special attention for co-operation and knowledge exchange between governmental organisations. Building Geo-data sets and enable re-use implies knowledge about its intended and future use, as where a lot of datasets now are offered in a one-solutions-use format. Last but not least policymakers should bear in mind that planning processes do imply a follow-up in realisation and monitoring.

FSC
www.fsc.org
MIX
Papier aus verantwortungsvollen Quellen
Paper from responsible sources
FSC® C105338

If you have any concerns about our products,
you can contact us on
ProductSafety@springernature.com

In case Publisher is established outside the EU,
the EU authorized representative is:
Springer Nature Customer Service Center GmbH
Europaplatz 3, 69115 Heidelberg, Germany

Printed by Libri Plureos GmbH
in Hamburg, Germany